# Universitext: Tracts in Mathematics

# Universitext

Editors (North America): S. Axler, F.W. Gehring, and P.R. Halmos

**Aksoy/Khamsi:** Nonstandard Methods in Fixed Point Theory
**Aupetit:** A Primer on Spectral Theory
**Booss/Bleecker:** Topology and Analysis
**Borkar:** Probability Theory; An Advanced Course
**Carleson/Gamelin:** Complex Dynamics
**Cecil:** Lie Sphere Geometry: With Applications to Submanifolds
**Chae:** Lebesgue Integration (2nd ed.)
**Charlap:** Bieberbach Groups and Flat Manifolds
**Chern:** Complex Manifolds Without Potential Theory
**Cohn:** A Classical Invitation to Algebraic Numbers and Class Fields
**Curtis:** Abstract Linear Algebra
**Curtis:** Matrix Groups
**DiBenedetto:** Degenerate Parabolic Equations
**Dimca:** Singularities and Topology of Hypersurfaces
**Edwards:** A Formal Background to Mathematics I a/b
**Edwards:** A Formal Background to Mathematics II a/b
**Foulds:** Graph Theory Applications
**Gardiner:** A First Course in Group Theory
**Gårding/Tambour:** Algebra for Computer Science
**Goldblatt:** Orthogonality and Spacetime Geometry
**Hahn:** Quadratic Algebras, Clifford Algebras, and Arithmetic Witt Groups
**Holmgren:** A First Course in Discrete Dynamical Systems
**Howe/Tan:** Non-Abelian Harmonic Analysis: Applications of $SL(2, R)$
**Howes:** Modern Analysis and Topology
**Humi/Miller:** Second Course in Ordinary Differential Equations
**Hurwitz/Kritikos:** Lectures on Number Theory
**Jennings:** Modern Geometry with Applications
**Jones/Morris/Pearson:** Abstract Algebra and Famous Impossibilities
**Kannan/Krueger:** Advanced Real Analysis
**Kelly/Matthews:** The Non-Euclidean Hyperbolic Plane
**Kostrikin:** Introduction to Algebra
**Luecking/Rubel:** Complex Analysis: A Functional Analysis Approach
**MacLane/Moerdijk:** Sheaves in Geometry and Logic
**Marcus:** Number Fields
**McCarthy:** Introduction to Arithmetical Functions
**Meyer:** Essential Mathematics for Applied Fields
**Mines/Richman/Ruitenburg:** A Course in Constructive Algebra
**Moise:** Introductory Problems Course in Analysis and Topology
**Morris:** Introduction to Game Theory
**Porter/Woods:** Extensions and Absolutes of Hausdorff Spaces
**Ramsay/Richtmyer:** Introduction to Hyperbolic Geometry
**Reisel:** Elementary Theory of Metric Spaces
**Rickart:** Natural Function Algebras
**Rotman:** Galois Theory
**Sagan:** Space-Filling Curves

*(continued after index)*

# Lennart Carleson   Theodore W. Gamelin

# Complex Dynamics

**With 28 Figures**

Springer

Lennart Carleson
Department of Mathematics
Royal Institute of Technology
S-100 44 Stockholm
Sweden
and Department of Mathematics
University of California
Los Angeles, CA 90024-1555
USA

Theodore W. Gamelin
Department of Mathematics
University of California
Los Angeles, CA 90024-1555
USA

*On the cover:* A filled-in Julia set with parabolic fixed point, attracting petals, and repelling arms.

Mathematics Subject Classification (1991): 30Cxx, 58Fxx

Library of Congress Cataloging-in-Publication Data
Carleson, Lennart.
    Complex dynamics/by L. Carleson and T. Gamelin.
      p.  cm. — (Universitext)
    Includes bibliographical references and index.
    ISBN 0-387-97942-5
    1. Functions of complex variables.   2. Mappings (Mathematics)
    3. Fixed point theory.   I. Title.
    QA331.7.C37   1993
    515′.9 — dc20                      92-32457

Printed on acid-free paper.

Production managed by Francine McNeill; manufacturing supervised by Vincent Scelta.
Photocomposed copy prepared from the author's $\mathcal{A}_{\mathcal{M}}\mathcal{S}$-TEX file.
Printed and bound by R.R. Donnelley & Sons, Harrisonburg, VA.
Printed in the United States of America.

9 8 7 6 5 4 3 2 (Corrected second printing, 1995)

ISBN 0-387-97942-5 Springer-Verlag New York Berlin Heidelberg
ISBN 3-540-97942-5 Springer-Verlag Berlin Heidelberg New York

# Preface

Complex dynamics is today very much a focus of interest. Though several fine expository articles were available, by P. Blanchard and by M.Yu. Lyubich in particular, until recently there was no single source where students could find the material with proofs. For anyone in our position, gathering and organizing the material required a great deal of work going through preprints and papers and in some cases even finding a proof. We hope that the results of our efforts will be of help to others who plan to learn about complex dynamics and perhaps even lecture. Meanwhile books in the field are beginning to appear. The Stony Brook course notes of J. Milnor were particularly welcome and useful. Still we hope that our special emphasis on the analytic side will satisfy a need.

This book is a revised and expanded version of notes based· on lectures of the first author at UCLA over several Winter Quarters, particularly 1986 and 1990. We owe Chris Bishop a great deal of gratitude for supervising the production of course notes, adding new material, and making computer pictures. We have used his computer pictures, and we will also refer to the attractive color graphics in the popular treatment of H.-O. Peitgen and P. Richter.

We have benefited from discussions with a number of colleagues, and from suggestions of students in both our courses. We would particularly like to acknowledge contributions from Peter Jones and M. Shishikura. Any reader familiar with the area will recognize the exposition of quasiconformal mappings from Ahlfors' book. It is often difficult to trace particular results to the rightful owners, particularly in such a rapidly developing area where so much seems to flow by word of mouth. We apologize for any inadequacy and for omissions.

L. Carleson and T.W. Gamelin
Los Angeles, March, 1992

# Contents

# I
# Conformal and Quasiconformal Mappings

We discuss some topics that are not included in the standard introductory graduate course in complex analysis. For the most part, we assume only the background provided by elementary graduate courses in real and complex analysis. There are a few exceptions. We use covering spaces and at one point the uniformization theorem, which can be found in Chapters 9 and 10 of [**A3**]. A readable discussion of the Poisson kernel and Fatou's theorem is given in Chapter 1 of [**Ho**]. In discussing the Beurling transform we appeal to the Calderón–Zygmund theorem, which can be found in [**St**] or [**A2**].

## 1.   Some Estimates on Conformal Mappings

A mapping $f$ is called *conformal* if it is analytic and one-to-one. Such mappings are also called *univalent*. Let $S$ be the collection of univalent functions in the open unit disk $\Delta = \{|z| < 1\}$ such that $f(0) = 0$ and $f'(0) = 1$. The compactness properties of this class are of crucial importance for our study.

THEOREM 1.1 (area theorem). *Let* $g(z) = 1/z + b_0 + b_1 z + \ldots$ *be univalent in* $\Delta$ *(with a pole at* $z = 0$*). Then* $\sum n|b_n|^2 \leq 1$.

*Proof.* For $0 < r < 1$ set $D_r = \mathbf{C} \backslash g(\Delta_r)$, where $\Delta_r = \{|z| < r\}$. By Green's theorem,

$$\text{area } D_r = \iint_{D_r} dx\, dy = \frac{1}{2i} \int_{\partial D_r} \bar{z}\, dz = \frac{-1}{2i} \int_{\partial \Delta_r} \bar{g}\, dg.$$

Substituting the power series expansions for $g$ and $g'$ and integrating, we obtain

$$\text{area } D_r = \pi \left( \frac{1}{r^2} - \sum_{n=1}^{\infty} n|b_n|^2 r^{2n} \right).$$

Since area $D_r \geq 0$, taking $r \to 1$ gives the result. $\square$

THEOREM 1.2. *If $f(z) = z + \sum_{n=2}^{\infty} a_n z^n \in \mathcal{S}$, then $|a_2| \leq 2$.*

*Proof.* Define $g(z) = 1/\sqrt{f(z^2)} = 1/z - a_2 z/2 + \cdots$. If $g(z_1) = g(z_2)$, then $f(z_1^2) = f(z_2^2)$, $z_1^2 = z_2^2$, and $z_1 = \pm z_2$. But $g$ is odd, so $z_1 = z_2$. Hence $g$ is univalent. The area theorem gives $|a_2| \leq 2$. $\square$

THEOREM 1.3 (Koebe one-quarter theorem). *If $f \in \mathcal{S}$, then the image of $f$ covers the open disk centered at 0 of radius one-quarter, that is, $f(\Delta) \supset \Delta(0, 1/4)$.*

*Proof.* Fix a point $c$, and suppose $f \neq c$ in $\Delta$. Then

$$\frac{cf(z)}{c - f(z)} = z + \left( a_2 + \frac{1}{c} \right) z^2 + \cdots$$

belongs to $\mathcal{S}$. Applying Theorem 1.2 twice, we obtain

$$\frac{1}{|c|} \leq |a_2| + \left| a_2 + \frac{1}{c} \right| \leq 2 + 2 = 4. \qquad \square$$

The Koebe function $f(z) = z/(1-z)^2 = \sum n z^n$ maps the disk $\Delta$ to the slit plane $\mathbf{C} \backslash (-\infty, -1/4]$. This shows that one-quarter is optimal.

The Koebe one-quarter theorem and the Schwarz lemma combine to give

$$\frac{1}{4} \leq \text{dist}(0, \partial f(\Delta)) \leq 1, \qquad f \in \mathcal{S},$$

where "dist" denotes distance. To prove the upper estimate, note that the image of $f \in \mathcal{S}$ cannot cover the closed unit disk, or else $f^{-1}$ would map $\Delta$ to a proper subdomain, contradicting $f'(0) = 1$. If we translate and scale, we obtain $\operatorname{dist}(f(z_0), \partial f(\Delta(\zeta_0, \delta))) \geq \delta |f'(z_0)|/4$ from the lower estimate whenever $f$ is univalent on the open disk $\Delta(z_0, \delta)$. This leads to the following.

THEOREM 1.4. *If $f$ is univalent on a domain $D$, and $z_0 \in D$, then*

$$\frac{1}{4}|f'(z_0)|\operatorname{dist}(z_0, \partial D) \leq \operatorname{dist}(f(z_0), \partial(f(D))) \leq 4|f'(z_0)|\operatorname{dist}(z_0, \partial D).$$

*Proof.* The lower estimate follows immediately from the statement preceding the theorem, and the upper estimate is obtained by applying the lower estimate to $f^{-1}$ at the point $f(z_0)$. $\square$

We aim now at proving some stronger distortion results.

THEOREM 1.5. *If $f \in \mathcal{S}$, then*

$$\left|\frac{zf''(z)}{f'(z)} - \frac{2|z|^2}{1-|z|^2}\right| \leq \frac{4|z|}{1-|z|^2}.$$

*Proof.* Fix $\zeta \in \Delta$ and consider

$$F(z) = \frac{f((z+\zeta)/(1+\bar{\zeta}z)) - f(\zeta)}{(1-|\zeta|^2)f'(\zeta)} = z + a_2(\zeta)z^2 + \cdots.$$

Then $F \in \mathcal{S}$, so by Theorem 1.2, $|a_2(\zeta)| \leq 2$. A computation shows

$$a_2(\zeta) = \frac{1}{2}\left\{(1-|\zeta|^2)\frac{f''(\zeta)}{f'(\zeta)} - 2\bar{\zeta}\right\}.$$

The assertion follows. $\square$

THEOREM 1.6 (distortion theorem). *If $f \in \mathcal{S}$, then*

$$\frac{1-|z|}{(1+|z|)^3} \leq |f'(z)| \leq \frac{1+|z|}{(1-|z|)^3}$$

*and*

$$\frac{|z|}{(1+|z|)^2} \leq |f(z)| \leq \frac{|z|}{(1-|z|)^2}.$$

*Proof.* Using Theorem 1.5, we estimate

$$\left|\frac{\partial}{\partial r}\log|f'(re^{i\theta})|\right| = \frac{1}{|z|}\operatorname{Re}\left\{\frac{zf''}{f'}\right\} \leq \frac{2|z|+4}{1-|z|^2}.$$

Integration from 0 to $r$ yields

$$\log|f'(re^{i\theta})| \leq \log\frac{1+|z|}{(1-|z|)^3},$$

which is the right-hand side of the first estimate. The right-hand side of the second estimate follows by another integration. To prove the lower bound for $|f'(z)|$, apply the upper bound to the function $F$ appearing in the proof of Theorem 1.5. The lower bound for $|f(z)|$ is obtained as follows. For fixed $0 < r < 1$, choose $z_0$ so that $|f(z_0)|$ is the minimum of $|f(z)|$ for $|z| = r$, and let $\gamma$ be the curve in $\Delta$ mapped by $f$ to the radial line segment from 0 to $f(z_0)$. Then $f'(\zeta)d\zeta$ has constant argument along $\gamma$, so that $|f(z_0)|$ coincides with the integral of $|f'(\zeta)||d\zeta|$ along $\gamma$. Integrating the lower estimate for $|f'|$, we obtain the lower estimate for $|f(z_0)|$. $\square$

THEOREM 1.7. *If $f \in S$, then*

$$\operatorname{dist}(f(z), \partial f(\Delta)) > \frac{1}{16}(1-|z|)^2, \qquad z \in \Delta.$$

*Proof.* We apply the Koebe one-quarter theorem (or the earlier distance estimate) to the function $F$ appearing in the proof of Theorem 1.5, to obtain

$$\operatorname{dist}(f(z), f(\partial\Delta)) \geq \frac{1}{4}|f'(z)|(1-|z|^2), \qquad z \in \Delta.$$

The lower estimate for $|f'(z)|$ in the distortion theorem then gives the result. $\square$

THEOREM 1.8. *If $f(z) = z + \sum_{n=2}^{\infty} a_n z^n \in S$, then $|a_n| < en^2$.*

*Proof.* Using the distortion theorem to estimate $|f(z)|$, we obtain

$$|a_n| = \left|\frac{1}{2\pi i}\int_{|z|=r}\frac{f(z)}{z^{n+1}}dz\right| \leq \frac{r^{1-n}}{(1-r)^2}, \qquad 0 < r < 1.$$

The choice $r = 1 - 1/n$ yields the desired estimate. $\square$

The Koebe function shows that each of the estimates of the distortion theorem are sharp. The lower estimate for $|f(z)|$ can be viewed as a strong form of the Koebe one-quarter theorem. The constant

1/16 appearing in Theorem 1.7 is also sharp. The coefficient estimate of Theorem 1.8 is rather crude, but good enough for our application in Section V.1. With some more effort we could obtain the estimate $|a_n| < en$ due to J.E. Littlewood (1925). It had been conjectured by L. Bieberbach (1916) that in fact $|a_n| \leq n$. The Bieberbach conjecture was settled affirmatively by L. deBranges (1985). The Koebe function shows this estimate is sharp.

Suppose $\mathcal{F}$ is a family of meromorphic functions in a domain $D \subset \overline{\mathbb{C}}$. We say $\mathcal{F}$ is a *normal family* if every sequence $\{f_n\}$ in $\mathcal{F}$ contains a subsequence that converges uniformly in the spherical metric, on compact subsets of $D$. By the Arzelà–Ascoli theorem, this is the same as saying the family is equicontinuous (with respect to the spherical metric) on every compact subset of $D$. Note that we allow $f_n \to \infty$ in the definition of normal family. Thus it is convenient to consider the function $f \equiv \infty$ to be meromorphic. The following theorem, first stated explicitly by P. Montel in his thesis (1907), will be used frequently.

THEOREM 1.9. *The family $\mathcal{F}$ of analytic functions on $D$ bounded by some fixed constant is normal.*

*Proof.* It is sufficient to prove the theorem for a disk (cover $D$ by disks), and we may take the unit disk. Then if $f \in \mathcal{F}$ satisfies $|f| \leq M$ and if $|z| < r < 1$, Cauchy's estimate implies $|f'(z)| \leq M/(1 - r)$. Thus the family is equicontinuous, hence normal. $\square$

THEOREM 1.10. *The family $\mathcal{S}$ is normal, and the limit of any sequence in $\mathcal{S}$ belongs to $\mathcal{S}$.*

*Proof.* Normality is clear since $|f(z)|$ is uniformly bounded on compact subsets of $\Delta$, by Theorem 1.6. Limit functions are in $\mathcal{S}$ on account of Hurwitz's theorem and the normalization $f'(0) = 1$. $\square$

## 2.   The Riemann Mapping

The Riemann mapping theorem asserts that if $D$ is a simply connected domain in $\overline{\mathbb{C}}$ whose boundary contains at least two points, there is a conformal mapping $\psi$ of the open unit disk $\Delta$ onto $D$. We can map 0 to any fixed point of $D$, and also specify the argument of

$\psi'(0)$, and then the Riemann mapping is unique. For the standard proof, see [A1]. We are concerned here with the boundary behavior of the Riemann mapping. We aim to establish two fundamental theorems due respectively to C. Carathéodory (1913) and E. Lindelöf (1915).

We will say that a compact set $K$ is *locally connected at* $z_0 \in K$ if for every sequence $\{z_n\} \subset K$ converging to $z_0$ there is, for $n$ large, a connected set $L_n \subset K$ containing both $z_0$ and $z_n$, such that $\text{diam}(L_n) \to 0$. A compact set is *locally connected* if it is locally connected at every point. On the right of Figure 1 is a connected set which is not locally connected.

**THEOREM 2.1** (Carathéodory). *Let $D$ be a simply connected domain in $\overline{\mathbb{C}}$ whose boundary has at least two points. Then $\partial D$ is locally connected if and only if the Riemann mapping $\psi : \Delta \to D$ extends continuously to the closed disk $\overline{\Delta}$.*

*Proof.* First suppose $\psi$ extends continuously. Let $z \in \partial D$, and let $\{z_n\}$ be a sequence in $\partial D$ converging to $z$. Taking $\zeta_n \in \psi^{-1}(z_n)$ and passing to a subsequence, we assume $\zeta_n \to \zeta$. By continuity, $\psi(\zeta) = z$. The arc $\gamma_n = \psi([\zeta_n, \zeta])$ is a connected subset of $\partial D$ containing $z_n$ and $z$. Since $\psi$ is uniformly continuous, $\text{diam}(\gamma_n) \to 0$. Hence $\partial D$ is locally connected at $z$.

For the other direction, suppose $\partial D$ is locally connected and assume $\infty = \psi(0) \in D$. Fix $\zeta_0 \in \partial\Delta$. For $\rho > 0$ consider $\gamma_\rho = \{\zeta \in \Delta : |\zeta - \zeta_0| = \rho\}$. Define

$$L(\rho) = \int_{\gamma_\rho} |\psi'(\zeta)||d\zeta|,$$

the length of the image curve $\psi(\gamma_\rho)$ in $D$. By the Cauchy-Schwarz inequality

$$L(\rho)^2 < \pi\rho \int_{\gamma_\rho} |\psi'(\zeta)|^2 |d\zeta|,$$

so

$$\int_0^\delta \frac{L(\rho)^2}{\rho} d\rho < \pi \iint_{\Delta \cap \Delta(\zeta_0,\delta)} |\psi'(\zeta)|^2 d\xi d\eta = \pi \text{area}\,\psi(\Delta \cap \Delta(\zeta_0,\delta)) < \infty.$$

Hence there is a sequence $\rho_n \to 0$ with $L(\rho_n) \to 0$. The curves $\Gamma_n = \psi(\gamma_{\rho_n})$ have endpoints $\alpha_n, \beta_n \in \partial D$ and $|\alpha_n - \beta_n| \to 0$. We

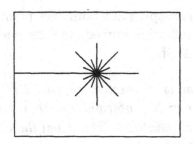 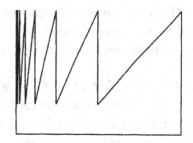

FIGURE 1. Locally connected, and not locally connected.

can assume that $\alpha_n, \beta_n$ tend to $w_0 \in \partial D$. Using the definition of local connectedness at $w_0$, we find connected subsets $L_n \subset \partial D$ with $w_n, \alpha_n, \beta_n \in L_n$ and $\operatorname{diam}(L_n) \to 0$. Now $\Gamma_n$ separates $D$ into two connected components, one containing $\infty = \psi(0)$. Let $D_n$ be the bounded component of $D \backslash \Gamma_n$. We claim that $D_n$ is contained in a bounded component of $\overline{\mathbf{C}} \backslash (\Gamma_n \cup L_n)$. Indeed, otherwise we could draw a simple arc from a fixed point $z_0 \in D_n$ to $\infty$ in $\overline{\mathbf{C}} \backslash (\Gamma_n \cup L_n)$, followed by another arc from $\infty$ to $z_0$ in $D$ crossing $\Gamma_n$ exactly once, to obtain a simple closed Jordan curve in $\overline{\mathbf{C}} \backslash L_n$ which separates $\alpha_n$ and $\beta_n$, contradicting the connectedness of $L_n$. Since $\operatorname{diam}(\Gamma_n \cup L_n) \to 0$, also $\operatorname{diam}(D_n) \to 0$, and hence $\psi$ is continuous at $\zeta_0$. $\square$

Let $D$ be the domain at the left of Figure 1, whose boundary includes a union of spikes emanating from 0 at rational angles, so that (say) the spike at angle $2\pi p/q$ has length $1/q$. Since $\partial D$ is locally connected, the Riemann mapping $\psi$ of $\Delta$ onto $D$ extends continuously to $\partial \Delta$. However the point $0 \in \partial D$ is the inverse image under $\psi$ of a Cantor set on the unit circle. Thus even a continuous Riemann mapping can have "bad " behavior.

Recall that a *Stolz angle* at $\zeta_0 \in \partial \Delta$ is a sector in $\Delta$ with vertex at $\zeta_0$ and aperture strictly less than $\pi$, bisected by the radius (see Figure 2). The property of Stolz angles we require is that if $u$ is a harmonic function, $u > 0$, such that $\liminf u(\zeta) \geq 1$ as $\zeta$ tends to a boundary arc on one side of $\zeta_0$ with endpoint at $\zeta_0$, then $\liminf u(\zeta) \geq \varepsilon > 0$ as $\zeta$ tends to $\zeta_0$ through any fixed Stolz angle. This is easiest to see in the upper half-plane, where a Stolz angle corresponds to a sector with vertex $x_0 \in \mathbf{R}$ of the form $\{\eta < \arg(z - x_0) < \pi - \eta\}$. Here one compares $u$ with the harmonic function $\arg(z - x_0)$ to obtain the estimate with $\varepsilon = \eta/\pi$.

For the local study of a conformal mapping at a boundary point $z_0$, we will appeal several times to the following immediate consequence of two well-known theorems of Lindelöf.

THEOREM 2.2. *Let $D \subset \overline{\mathbf{C}}$ be simply connected, suppose $\partial D$ has more than one point, and let $\psi(\zeta)$ map $\Delta$ conformally to $D$. Let $\gamma$ be a Jordan arc in $D$ except for one endpoint $z_0 \in \partial D$. Then the curve $\psi^{-1} \circ \gamma$ terminates in a point $\zeta_0 \in \partial\Delta$, and $\psi(\zeta) \to z_0$ as $\zeta \to \zeta_0$ inside any Stolz angle at $\zeta_0$.*

*Proof.* Let $\Gamma$ be the curve $\psi^{-1}(\gamma \backslash \{z_0\})$ in $\Delta$, which clusters on $\partial\Delta$. We first want to prove that $\Gamma$ has a unique cluster point $\zeta_0$ on $\partial\Delta$. For this we could use Fatou's theorem, but let us give a more geometric proof. We wish to use some conformal invariant and let us consider a Dirichlet integral. We may assume $D$ is bounded. Define

$$f(z) = \left( \log^+ \frac{1}{|z - z_0| + \varepsilon} \right)^{1/3}.$$

A simple computation shows that independently of $\varepsilon > 0$,

$$D(f) = \iint_D |\nabla f|^2 dx dy \le M < \infty.$$

Let $F(\zeta) = f(\psi(\zeta))$ so that

$$D(F) = \iint_\Delta |\nabla F|^2 d\xi d\eta = D(f) \le M.$$

If $\Gamma$ does not tend to a point $\zeta_0$, there exists a sector $\theta_1 < \theta < \theta_2$ such that $\Gamma$ crosses to opposite sides of the sector infinitely often as it tends to the boundary. Then

$$\int_0^1 \left| \frac{\partial F}{\partial r}(re^{i\theta}) \right| dr \ge \left( \log \frac{1}{\varepsilon} \right)^{1/3} - C, \qquad (C = F(0)),$$

for $\theta_1 < \theta < \theta_2$. Using the Schwarz inequality we find

$$\int_{\theta_1}^{\theta_2} d\theta \int_0^1 \left| \frac{\partial F}{\partial r}(re^{i\theta}) \right|^2 r dr \ge \delta \left( \log \frac{1}{\varepsilon} \right)^{2/3} - 2\pi C,$$

which contradicts $D(F) \le M$. Thus $\Gamma$ accumulates at only one point $\zeta_0$ of the circle, and we can assume that $\zeta_0 = 1$.

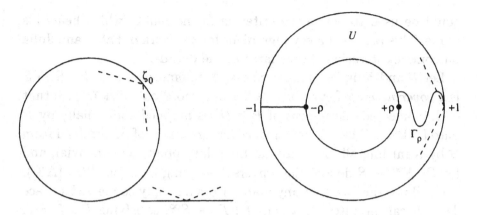

FIGURE 2. Stolz angles.

For $\rho < 1$ let $\Gamma_\rho$ be the subarc of $\Gamma$ lying in the annulus $\rho < |\zeta| < 1$ with one endpoint $\zeta_\rho$ on the circle $|\zeta| = \rho$ and the other at the terminal point 1 of $\Gamma$. Then $\Gamma_\rho$ divides the slit annulus $\{\rho < |\zeta| < 1\}\backslash(-1, -\rho)$ into two components. Let $U$ be (say) the top component. Let $\omega(\zeta)$ be harmonic in $U$, equal to 1 on $\Gamma_\rho$ and equal to 0 on the rest of $\partial U$. Let $\varepsilon = \varepsilon(\rho)$ be the supremum of $|\psi(\zeta) - z_0|$ over $\zeta \in \Gamma_\rho$, so that $\varepsilon \to 0$ as $\rho \to 1$. Assume as we may that $D \subset \{|z| < 1/2\}$ so that $|\psi(\zeta) - z_0| \leq 1$. By the maximum principle and the fact that $\log |\psi(\zeta) - z_0| \leq 0$ on $U$,

$$\log |\psi(\zeta) - z_0| < \omega(\zeta) \log \varepsilon$$

in $U$. We obtain a minorant $\omega^*(\zeta)$ of $\omega(\zeta)$ by replacing $\Gamma_\rho$ by the arc $A$ from 1 to $-1$ along the bottom half of $\{|\zeta| = 1\}$ and defining $\omega^*(\zeta)$ harmonic in the ring $R = \{\rho < |\zeta| < 1\}$ slit along $(-1, -\rho)$. Then $\omega^*(\zeta) = 0$ on $\partial R$ except on $A$ where $\omega^* = 1$. The estimate above now holds in $U$, with $\omega^*$ replacing $\omega$. According to our preliminary remarks, $\liminf \omega^*(\zeta) \geq c > 0$ as $\zeta \to 1$ inside any fixed Stolz angle $S$. Hence

$$\limsup_{S \ni \zeta \to 1} |\psi(\zeta) - z_0| \leq \varepsilon^c,$$

so that $\psi(\zeta) \to z_0$ as asserted. $\square$

## 3.   Montel's Theorem

It was P. Montel (1911) who formulated the notion of a normal family of meromorphic functions and realized that the modular function could be used to derive the criterion for normality which bears his name. This provided a key ingredient for the work of Fatou and Julia on complex iteration theory later in the decade.

Let $R$ and $S$ be Riemann surfaces. A holomorphic map $P : R \to S$ is a *covering map* if every $w \in S$ lies in a coordinate disk $U$ such that each connected component of $P^{-1}(U)$ is mapped conformally by $P$ onto $U$. Let $S^\infty$ be the universal covering surface of $S$, obtained from $S$ by regarding all loops around boundary points as nontrivial, and let $P : S^\infty \to S$ denote the universal covering map (see **[Sp]**,**[A3]**). We will use the fact that any map $f$ from a simply connected surface $D$ to $S$ can be lifted to a map $\tilde{f} : D \to S^\infty$, satisfying $P \circ \tilde{f} = f$. The value $\tilde{f}(z_0) \in P^{-1}(f(z_0))$ can be specified arbitrarily, and this determines the lift $f$ uniquely. The other lifts of $f$ are the maps $\varphi \circ \tilde{f}$, where $\varphi$ is a covering transformation, that is, a conformal self-map of $S^\infty$ satisfying $P \circ \varphi = P$. One consequence of the lifting property is that if $f$ is itself a covering map of a simply connected surface $D$ onto $S$, then the lift $\tilde{f} : D \to S^\infty$ is one-to-one and onto, so that $D$ is conformally equivalent to $S^\infty$.

THEOREM 3.1. *If $D = \mathbf{C} \backslash \{0, 1\}$ is the thrice-punctured sphere, then $D^\infty$ is conformally equivalent to the open unit disk $\Delta$.*

*Proof.* Since $\Delta$ is conformally equivalent to the upper half-plane $\mathbf{H} = \{y > 0\}$, it suffices to find a covering map of $\mathbf{H}$ onto $D$. We construct such a map, the so-called *modular function*, as follows. Let $E = \{z : 0 < x < 1, |z - 1/2| > 1/2\}$. In view of the Riemann mapping theorem, there is $\psi$ mapping $E$ to $\mathbf{H}$ fixing $0, 1, \infty$. Let $E^*$ denote the reflection of $E$ through the circle $\{|z - 1/2| = 1/2\}$. By the Schwarz reflection principle, we can extend $\psi$ to a conformal mapping of $E \cup E^*$ to $\mathbf{C} \backslash (-\infty, 0] \cup [1, \infty)$. By continuing to reflect we can extend $\psi$ to all of $\{0 < x < 1, y > 0\}$ taking its values in $\mathbf{C} \backslash \{0, 1\}$. By reflecting across the vertical lines $\{x = n\}$ for $n$ an integer, we can extend $\psi$ to all of $\mathbf{H}$. From the construction one sees that the extended $\psi$ is a covering map of $\mathbf{H}$ over $D$. □

THEOREM 3.2 (Montel's theorem). *Let $\mathcal{F}$ be a family of meromorphic functions on a domain $D$. If there are three fixed values that are omitted by every $f \in \mathcal{F}$, then $\mathcal{F}$ is a normal family.*

*Proof.* We may assume that $D$ is a disk, and, by composing with a

Möbius transformation, we may assume that the functions in $\mathcal{F}$ do not assume the values $0, 1, \infty$. Let $S = \mathbf{C}\backslash\{0, 1\}$. By Theorem 3.1, there is a covering map $\psi : \Delta \to S$. Let $\tilde{f} : D \to \Delta$ be a lift of $f \in \mathcal{F}$, so that $\tilde{f} \circ \psi = f$. Then $\{\tilde{f} : f \in \mathcal{F}\}$ is a normal family, by Theorem 1.9, and this implies that $\mathcal{F}$ is normal. $\square$

The uniformization theorem states that every simply connected Riemann surface is conformally equivalent to exactly one of the unit disk $\Delta$, the complex plane $\mathbf{C}$, or the Riemann sphere $\overline{\mathbf{C}}$. Moreover, every Riemann surface has one of these three as its universal cover. The three cases are referred to as *hyperbolic*, *parabolic*, and *elliptic*, respectively. For a proof of the uniformization theorem see [**A3**]. We will require the following criterion for a surface to be hyperbolic.

THEOREM 3.3. *Let $R$ be a Riemann surface that has a nonconstant meromorphic function omitting at least three values. Then $R^\infty$, its universal covering surface, is conformally equivalent to the unit disk, with any given point corresponding to the origin.*

*Proof.* Let $P : R^\infty \to R$ be the universal covering map. If $f$ on $R$ omits three values, then $f \circ P$ is a nonconstant meromorphic function on $R^\infty$ that omits three values, say $\{0, 1, \infty\}$. Using a covering map of $\Delta$ onto $\mathbf{C}\backslash\{0, 1\}$, we can lift $f \circ P$ to a nonconstant function from $R^\infty$ to the the open unit disk $\Delta$. Thus $R^\infty$ is not conformally equivalent to $\overline{\mathbf{C}}$ or $\mathbf{C}$, and consequently $R$ is hyperbolic. $\square$

## 4.   The Hyperbolic Metric

A conformal mapping of $\Delta$ onto itself has the form $w = e^{i\theta}(z - a)/(1 - \bar{a}z)$ for some $0 \le \theta \le 2\pi$ and $|a| < 1$. A computation yields

$$\left|\frac{dw}{dz}\right| = \frac{1 - |w|^2}{1 - |z|^2},$$

from which we see that

$$d\rho = \frac{2|dz|}{1 - |z|^2} = \frac{2|dw|}{1 - |w|^2}$$

is invariant under the mapping. The metric $d\rho$ is called the *hyperbolic* (or *Poincaré*) metric on $\Delta$. From the hyperbolic metric, we get a distance $\rho(z_1, z_2)$ in the obvious way by integrating along curves from $z_1$ to $z_2$ and taking the infimum. One checks that

$$\rho(0, z) = \log \frac{1 + |z|}{1 - |z|}, \qquad z \in \Delta.$$

Using a covering map $P : \Delta \to S$, we define a hyperbolic metric $d\rho_S$ on any hyperbolic Riemann surface $S$ by declaring that $P$ induces an isometry at every point. In other words, we set

$$d\rho_S(z) = \frac{2|dw|}{1 - |w|^2}, \qquad z = P(w) \in S.$$

If $\varphi$ is a local determination for $P^{-1}$, this formula becomes

$$d\rho_S(z) = \frac{2|\varphi'(z)|}{1 - |\varphi(z)|^2} |dz|, \qquad z \in S.$$

Since $d\rho$ is invariant under conformal self-maps of $\Delta$, this is independent of the choice of the branch $\varphi$ of $P^{-1}$, and it is also independent of the choice of the covering map $P$. As an example, using the conformal map $\varphi(z) = (z - i)/(z + i)$ of the upper half-plane $\mathbf{H}$ onto the unit disk $\Delta$, we compute the hyperbolic metric of the upper half-plane to be

$$d\rho_{\mathbf{H}}(z) = \frac{|dz|}{y}, \qquad z = x + iy, y > 0.$$

A holomorphic map $f : R \to S$ allows us to pull back differentials on $S$ to differentials on $R$ by the obvious formula (chain rule). Thus we can pull back the hyperbolic metric $d\rho_S$ on $S$ to a metric on $R$, which we denote by $f^*(d\rho_S)$.

THEOREM 4.1. *Suppose $f$ maps a hyperbolic Riemann surface $R$ holomorphically into a hyperbolic surface $S$. Then*

$$\begin{aligned} f^*(d\rho_S) &\leq d\rho_R, \\ \rho_S(f(z_1), f(z_2)) &\leq \rho_R(z_1, z_2), \qquad z_1, z_2 \in R, \end{aligned}$$

*with strict inequality unless $f$ lifts to a Möbius transformation mapping $\Delta$ onto $\Delta$.*

*Proof.* The second estimate is just the integrated version of the first. In the case $R = S = \Delta$, the first boils down to G. Pick's (1915) invariant form of the Schwarz lemma,

$$|f'(z)| \leq \frac{1 - |f(z)|^2}{1 - |z|^2}, \qquad z \in \Delta.$$

This is obtained from the Schwarz lemma by pre- and postcomposing with the appropriate Möbius transformations. For the general case, we check the estimate at a fixed point $z_0 \in R$. Let $w_0 = f(z_0)$, and consider covering maps $P : \Delta \to R$ and $Q : \Delta \to S$ satisfying $P(0) = z_0$, $Q(0) = w_0$. We lift $f$ to an analytic function $F : \Delta \to \Delta$ satisfying $Q \circ F = f \circ P$, $F(0) = 0$. It suffices to check that $F^*(d\rho(\zeta)) \leq d\rho(z)$ at $z = 0$. Now

$$F^*(d\rho(\zeta)) = \frac{2|F'(z)|}{1 - |F(z)|^2}|dz|, \qquad z \in \Delta,$$

so the estimate becomes $|F'(0)| \leq 1$. This holds, with equality only when $F$ maps $\Delta$ conformally onto $\Delta$. $\square$

THEOREM 4.2. *Suppose $R \subseteq S$. Then $d\rho_R \geq d\rho_S$ with strict inequality unless $R = S$.*

*Proof.* Apply Theorem 4.1 to the inclusion map. $\square$

THEOREM 4.3. *Let $D$ be a domain in $\mathbf{C}$, and for $z \in D$, let $\delta(z)$ denote the distance from $z$ to $\partial D$. Then if $D$ is simply connected,*

$$\frac{1}{2}\frac{|dz|}{\delta(z)} \leq d\rho_D(z) \leq 2\frac{|dz|}{\delta(z)}. \tag{4.1}$$

*For general domains as $z \to \partial D$,*

$$\frac{1 + o(1)}{\delta(z)\log(1/\delta(z))}|dz| \leq d\rho_D(z) \leq 2\frac{|dz|}{\delta(z)}. \tag{4.2}$$

*Proof.* For the right-hand sides, observe that $D' = \Delta(z_0, \delta(z_0)) \subset D$, so by Theorem 4.2, $d\rho_D(z_0) \leq d\rho_{D'}(z_0) = 2|dz|/\delta(z_0)$. For the the left-hand side of (4.1), let $\psi$ map $\Delta$ to $D$ with $\psi(0) = z_0$. By the Koebe one-quarter theorem, $\psi$ takes all values in $\Delta(z_0, |\psi'(0)|/4)$, and hence $\delta(z_0) \geq |\psi'(0)|/4$, which gives the desired inequality.

To prove the left-hand side of (4.2), let $z_1$ be a point in $\partial D$ closest to $z_0$ and choose two other points $z_2$ and $z_3 \in \partial D$. Assume $z_1 = 0$, $z_2 = 1$, and $z_3 = \infty$. Let $D'$ be the complement of $\{0, 1, \infty\}$. Then $d\rho_D \geq d\rho_{D'}$, so we only have to estimate $d\rho_{D'}$ near $z_1 = 0$. Let $\psi : \mathbf{H} \to D'$ be the modular function, constructed in the proof of Theorem 2.1. Then $\psi(w + 2) = \psi(w)$, so that $\psi$ can be expanded as a Laurent series in $e^{i\pi w}$. Since $\psi$ maps a neighborhood of $i\infty$ in the strip $\{0 \leq \mathrm{Re}\, w < 2\}$ onto a punctured neighborhood of $\infty$, $\psi$ has a simple pole, and $z(w) = 1/\psi(w)$ has a simple zero,

$$z(w) = b_1 e^{i\pi w} + b_2 e^{2i\pi w} + \cdots, \qquad \mathrm{Im}\, w > 0,$$

where $b_1 \neq 0$. Solving for $w$, we obtain

$$w(z) = \frac{1}{\pi i} \log z + \text{analytic}, \qquad \frac{dw}{dz} = \frac{1}{\pi i z} + \text{analytic}.$$

Now $z(w)$ is a covering map of $\mathbf{H}$ over $D'$, so it can be used to express the hyperbolic metric of $D'$ in terms of that of $\mathbf{H}$,

$$d\rho_{D'}(z) = \frac{|dw|}{\mathrm{Im}\, w} = \left|\frac{dw}{dz}\right| \frac{|dz|}{\mathrm{Im}\, w(z)} = \frac{|dz|}{|z| \log(1/|z|)} \left[1 + \frac{O(1)}{\log(1/|z|)}\right],$$

from which the estimate follows. $\square$

We end this section with an estimate for Green's function which will be useful for the study of the Mandelbrot set. The estimate, which appears in [CaJ], is a simple variant of estimates for harmonic measure and Green's function in the thesis of A. Beurling (pp. 29-30 of [Beu]).

THEOREM 4.4. *Let $D$ be simply connected, and let $\delta(z)$ be the distance from $z \in D$ to $\partial D$. Let $G(z, z_0)$ be Green's function for $D$ with pole at $z_0 \in D$. If $z \in D$ satisfies $G(z, z_0) \leq 1$, then there is an arc $\gamma$ from $z_0$ to $z$ in $D$ such that*

$$G(z, z_0) \leq 3 \exp\left(-\frac{1}{2} \int_\gamma \frac{|dw|}{\delta(w)}\right).$$

*Proof.* Map $D$ conformally to the unit disk $\Delta$ with $z_0$ going to 0. In these coordinates $G(z, z_0) = -\log|\zeta(z)|$. Also note the identity

$$G(z, z_0) = \log \frac{1}{|\zeta|} = \log \frac{1 + e^{-A}}{1 - e^{-A}}, \qquad (4.3)$$

where

$$A = 2 \int_0^{|\zeta|} \frac{dr}{1 - r^2} = \log \frac{1 + |\zeta|}{1 - |\zeta|}$$

is the hyperbolic distance from 0 to $\zeta$, that is, from $z_0$ to $z$. From (4.3) and the assumption $G(z, z_0) \leq 1$, we obtain $e^{-A} \leq (e - 1)/(e + 1) \leq 1/2$. Hence

$$\log \frac{1 + e^{-A}}{1 - e^{-A}} \leq 3e^{-A}.$$

By Theorem 4.3,

$$A \geq \frac{1}{2} \int_\gamma \frac{|dw|}{\delta(w)},$$

where $\gamma$ is the (geodesic) path in $D$ corresponding to the ray in $\Delta$ from 0 to $\zeta$. If we insert these estimates in (4.3) the theorem follows. $\square$

## 5. Quasiconformal Mappings

The reason why quasiconformal mappings are so useful in iteration theory is that these mappings allow for the kind of complicated behavior that arises and at the same time admit a way to connect analytic information to geometric. The person who first realized the power of quasiconformal mappings in dynamical systems was D. Sullivan (1985), and it is now a standard tool. Actually, it is the main new idea in the field.

Suppose $f$ has a continuous first derivative, that is, $f \in C^1$. We will use the standard notation $dz = dx + idy$, $d\bar{z} = dx - idy$ and

$$f_z = \tfrac{1}{2}(f_x - if_y),$$
$$f_{\bar{z}} = \tfrac{1}{2}(f_x + if_y).$$

If $w = f(z)$, we also write

$$df = dw = f_z dz + f_{\bar{z}} d\bar{z}.$$

The Jacobian $J_f$ of $f$ is given by

$$J_f = |f_z|^2 - |f_{\bar{z}}|^2.$$

Thus $f$ preserves orientation if and only if $|f_{\bar{z}}| < |f_z|$. We are concerned only with mappings that satisfy this condition.

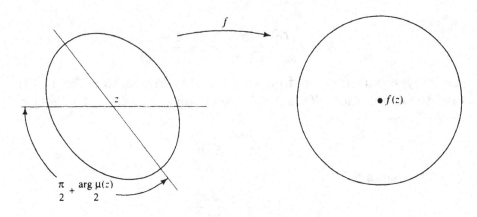

FIGURE 3. Ellipses to circles.

We define the (*complex*) *dilatation* $\mu = \mu_f$ of $f$ by

$$\mu = \frac{f_{\bar{z}}}{f_z}.$$

The dilatation $\mu$ is also called the *Beltrami coefficient* of $f$, and the equation $f_{\bar{z}} = \mu f_z$ is the *Beltrami equation*. Note that $|\mu| < 1$ if $f$ preserves orientation and that $\mu = 0$ if and only if $f$ is conformal. Since $f$ is differentiable in $D$ we can associate to $f$ an infinitesimal ellipse field in $D$ by assigning to each $z \in D$ the ellipse that is mapped to a circle by $f$. The argument of the major axis of this infinitesimal ellipse is $\pi/2 + \arg(\mu)/2$, and the ratio of minor to major axis is $(|f_z| - |f_{\bar{z}}|)/(|f_z| + |f_{\bar{z}}|) = (1 - |\mu|)/(1 + |\mu|)$.

Motivated by these relations we associate to any $\mu$ satisfying $|\mu| < 1$ an infinitesimal ellipse field, that is, a choice of direction and eccentricity at each point. The argument of the major axis is $\pi/2 + \arg(\mu)/2$, and the axis ratio is $(1 - |\mu|)/(1 + |\mu|)$. Solving the Beltrami equation $f_{\bar{z}} = \mu f_z$ is then equivalent to finding $f$ whose associated ellipse field coincides with that of $\mu$.

A smooth map $f$ is *k-quasiconformal* in $D$ if its dilatation $\mu$ satisfies $|\mu| \le k < 1$. For most applications we will have $D = \overline{\mathbf{C}}$, and then we just say $f$ is *k-quasiconformal*. Geometrically, the condition means that there is a fixed bound on the stretching of $f$ in any given direction compared to any other direction. One clearly has

$$|\mu_{f^{-1}}(f(z))| = |\mu_f(z)|.$$

In fact, the ellipse corresponding to $f^{-1}$ at $f(z)$ is the rotate by 90°

of the ellipse corresponding to $f$ at $z$. If $\varphi$ is conformal, then

$$\mu_{\varphi \circ f}(z) = \mu_f(z),$$
$$\mu_{f \circ \varphi}(z) = \mu_f(\varphi(z))\overline{\varphi'(z)}/\varphi'(z).$$

Note that postcomposing by a conformal map does not change $\mu$. This is just what we expect, since such a map does not alter which ellipse gets mapped to a circle. On the other hand, precomposing by a conformal map can change the direction but not the eccentricity of such an ellipse, as expressed in the second equation above.

If $f$ and $g$ are smooth quasiconformal homeomorphisms of the sphere $\overline{\mathbf{C}}$ whose Beltrami coefficients coincide, then $g = \varphi \circ f$ for some Möbius function $\varphi$. Indeed $g \circ f^{-1}$ maps infinitesimal circles to infinitesimal circles, hence $g \circ f^{-1}$ is a conformal self-map of $\overline{\mathbf{C}}$.

Suppose $f$ is $k$-quasiconformal and $g$ is $k'$-quasiconformal, and set

$$K = \frac{1+k}{1-k}, \qquad K' = \frac{1+k'}{1-k'}.$$

Then the $K''$ corresponding to $f \circ g$ is at most $K \cdot K'$ (since the axis ratios of the ellipses at most multiply) and so $f \circ g$ is $k''$-quasiconformal for some $k'' = k''(k, k')$.

## 6.    Singular Integral Operators

We will make repeated use of the Cauchy–Green formula for a bounded domain $D$ and a smooth function $f$,

$$f(z) = \frac{1}{2\pi i} \int_{\partial D} \frac{f(\zeta)}{\zeta - z} d\zeta - \frac{1}{\pi} \iint_D \frac{f_{\bar{z}}(\zeta)}{\zeta - z} d\xi d\eta, \qquad z \in D, \; \zeta = \xi + i\eta.$$

Let $\varphi \in L^1$ have compact support and define

$$T\varphi(z) = -\frac{1}{\pi} \iint_{\mathbf{C}} \frac{\varphi(\zeta)}{\zeta - z} d\xi d\eta.$$

The operator $T$ is a convolution operator, whose kernel $1/z$ belongs to $L^1$ on any bounded set. Consequently $T\varphi$ is locally integrable. From the formula one sees that $T\varphi$ is analytic on any open set on which $\varphi = 0$ and at $\infty$, and $(T\varphi)(\infty) = 0$. Applying the Cauchy–Green formula for a large disk to a test function $\psi \in C_c^\infty$, we obtain the identity

$$\iint \psi_{\bar{z}}(z)(T\varphi)(z)dxdy = -\iint \psi(\zeta)\varphi(\zeta)d\xi d\eta,$$

which means that

$$\frac{\partial}{\partial \bar{z}}T\varphi = \varphi$$

in the sense of distributions.

For smooth $\varphi$ with compact support we define the *Beurling transform* $S\varphi$ of $\varphi$ by

$$S\varphi = (T\varphi)_z\,.$$

Thus $S(f_{\bar{z}}) = f_z$ whenever $f$ is a smooth function which is analytic at $\infty$ and vanishes there. By differentiating we see that $S\varphi$ is given formally by

$$-\frac{1}{\pi}\iint \frac{\varphi(\zeta)}{(\zeta - z)^2}d\xi d\eta.$$

This is a singular integral operator, and the integral diverges absolutely. To make this rigorous we fix $z_0$ and define $\psi(z) = T\varphi(z) - \varphi(z_0)\bar{z}$. The Cauchy–Green formula yields

$$\bar{z} = -\frac{1}{\pi}\iint\limits_{|\zeta| < R} \frac{d\xi d\eta}{\zeta - z} = T\chi_R(z), \qquad |z| < R,$$

where $\chi_R$ is the characteristic function of the disk $\{|z| < R\}$. If $R$ is chosen so large that $\varphi = 0$ off $\Delta(0, R)$, this implies for $|z_0| < R$ and $h$ small that

$$\frac{\psi(z_0 + h) - \psi(z_0)}{h} = -\frac{1}{\pi}\iint\limits_{|\zeta| < R} \frac{\varphi(\zeta) - \varphi(z_0)}{(\zeta - z_0)(\zeta - z_0 - h)}d\xi\,d\eta.$$

Since $\psi_z = S\varphi$, we obtain in the limit as $h \to 0$

$$S\varphi(z_0) = -\frac{1}{\pi}\iint\limits_{|\zeta| < R} \frac{\varphi(\zeta) - \varphi(z_0)}{(\zeta - z_0)^2}d\xi\,d\eta.$$

In view of the identity

$$\iint\limits_{a < |\zeta| < b} \frac{d\xi d\eta}{\zeta^2} = 0,$$

we obtain an expression for $S$ as a principal value integral,

$$(S\varphi)(z) = \lim_{\varepsilon \to 0} -\frac{1}{\pi} \iint_{\varepsilon < |\zeta| < 1/\varepsilon} \frac{\varphi(\zeta)}{(\zeta - z)^2} d\xi \, d\eta.$$

The limit in this formula exists if $\varphi$ is Hölder continuous. The Beurling transform $S$ is bounded in the usual $L^p$ norms $1 < p < \infty$. More precisely,

$$\|S(\varphi)\|_2 = \|\varphi\|_2 ,$$
$$\|S(\varphi)\|_p \leq K_p \|\varphi\|_p , \qquad 1 < p < \infty,$$

where $K_p$ depends continuously on $p$, and in particular $K_p \to 1$ as $p \to 2$. These estimates allow us to extend $S$ to all of $L^p$, for $1 < p < \infty$. The usual convolution argument with an approximate identity shows that if a distribution $f$ has distribution derivatives $f_z$ and $f_{\bar{z}}$ in $L^p$ for some $1 < p < \infty$, then (the extended) $S$ satisfies $S(f_{\bar{z}}) = f_z$. A proof of these facts is included in Ahlfors' exposition [A2] of quasiconformal mapping. For a detailed exposition of singular integral operators, see [St].

## 7.   The Beltrami Equation

Fix $0 < k < 1$, and let $L^\infty(k, R)$ denote the measurable functions on $\mathbf{C}$ bounded by $k$ and supported in $\Delta(0, R)$. We let $QC^1(k, R)$ denote the continuously differentiable homeomorphisms $f$ of $\mathbf{C}$ such that $f_{\bar{z}} = \mu f_z$ for some $\mu \in L^\infty(k, R)$, normalized so that $f(z) = z + \mathcal{O}(1/z)$ as $z \to \infty$. Note that $f$ is analytic for $|z| > R$.

LEMMA 7.1. *If $f \in QC^1(k, R)$, then the inverse function $f^{-1}$ belongs to $QC^1(k, 4R)$.*

*Proof.* This follows directly from the Koebe one-quarter theorem and the formula for the Beltrami coefficient of $f^{-1}$ given in Section 5. $\square$

Let $f \in QC^1(k, R)$, and consider the Cauchy–Green formula for a disk,

$$f(z) = \frac{1}{2\pi i} \int_{|\zeta| = r} \frac{f(\zeta)}{\zeta - z} d\zeta - \frac{1}{\pi} \iint_{|\zeta| < r} \frac{f_{\bar{z}}(\zeta)}{\zeta - z} d\xi d\eta, \qquad |z| < r.$$

For $r > R$ the $\mathcal{O}(1/z)$ term does not contribute to the first integral. We can replace $f(\zeta)$ by $\zeta$ in the first integral, and this leads to

$$f(z) - z = -\frac{1}{\pi} \iint_{|\zeta| < R} \frac{f_{\bar{z}}(\zeta)}{\zeta - z} d\xi d\eta = Tf_{\bar{z}}(z). \qquad (7.1)$$

Thus if we set $g = f_z - 1$ and use $f_{\bar{z}} = \mu f_z$, we obtain

$$g = S(f_{\bar{z}}) = S(\mu f_z) = S(\mu g) + S(\mu).$$

In terms of the operator

$$U_\mu(g) = S(\mu g), \qquad g \in L^p,$$

this equation becomes

$$(I - U_\mu)g = S(\mu).$$

The operator $U_\mu$ has $L^p$-norm $\|U_\mu\|_p \leq kK_p < 1$ for $p > 2$ sufficiently close to 2. We fix $p = p(k) > 2$, once and for all, so that

$$kK_p < 1.$$

Then $\|U_\mu\| < 1$ and $I - U_\mu$ is invertible, with inverse

$$(I - U_\mu)^{-1} = \sum_{n=0}^{\infty} U_\mu^n,$$

$$\|(I - U_\mu)^{-1}\| \leq \sum_{n=0}^{\infty} (kK_p)^n = \frac{1}{1 - kK_p}.$$

Thus we can solve for $g$, to obtain

$$g = (I - U_\mu)^{-1} S(\mu) \in L^p. \qquad (7.2)$$

This leads to $L^p$-estimates for $g$,

$$\|g\|_p \leq \|(I - U_\mu)^{-1}\|_p \|S\|_p \|\mu\|_p \leq \frac{(\pi R^2)^{1/p}}{1 - kK_p}. \qquad (7.3)$$

Since the Jacobian of $f$ is given by

$$J_f = |f_z|^2 - |f_{\bar{z}}|^2 = (1 - |\mu|^2)|f_z|^2,$$

$J_f$ belongs to $L^{p/2}$ on bounded sets. This has the following consequence.

**THEOREM 7.2.** *Let* $p = p(k)$ *be as above. If* $f \in QC^1(k, R)$ *and*
$E \subset \Delta(0, R)$, *then*

$$\text{area } f(E) \leq C(\text{area } E)^{1-2/p},$$
$$\text{area } E \leq C(\text{area } f(E))^{1-2/p},$$

*where* $C$ *depends only on* $k$ *and* $R$.

*Proof.* By Hölder's inequality applied with exponent $p/2$,

$$\text{area } f(E) = \iint_E J_f dx dy \leq \left( \iint_E |f_z|^p dx dy \right)^{2/p} (\text{area } E)^{1-2/p}.$$

Since $f_z = g + 1$, the $L^p$-estimate (7.3) for $g$ gives the first estimate.
Lemma 7.1 permits us to apply the first estimate to $f^{-1}$, and this
yields the second with a larger constant. $\square$

We can also prove a uniform estimate on the modulus of continuity
of $f$.

**THEOREM 7.3.** *Let* $p = p(k)$ *be as above. If* $f \in QC^1(k, R)$, *then*

$$|f(z_1) - f(z_2)| \leq C|z_1 - z_2|^{1-2/p}, \qquad |z_1|, |z_2| < R,$$

*where* $C$ *depends only on* $k$ *and* $R$.

*Proof.* We use the formula (7.1) and find

$$|f(z_1) - f(z_2)| \leq |z_1 - z_2| + \frac{1}{\pi} \|f_{\bar{z}}\|_p \left( \iint \frac{|z_1 - z_2|^q}{|\zeta - z_1|^q |\zeta - z_2|^q} d\xi \, d\eta \right)^{1/q},$$

where $q = p/(p-1)$ is the conjugate index of $p$. We split the integral
into three pieces $D_1 = \{|\zeta - z_1| < \frac{1}{2}|z_1 - z_2|\}$, $D_2 = \{|\zeta - z_2| < \frac{1}{2}|z_1 - z_2|\}$, and $D_3 = \mathbf{C} \setminus (D_1 \cup D_2)$. Since $|\zeta - z_2| \geq |z_1 - z_2|/2$ on
$D_1$, the integral over $D_1$ is controlled by

$$\iint_{D_1} \frac{d\xi \, d\eta}{|\zeta - z_1|^q} \leq \iint_{0 \leq r \leq |z_1 - z_2|} \frac{r \, dr \, d\theta}{r^q} = c|z_1 - z_2|^{2-q}.$$

The integral over $D_2$ is treated similarly. The integral over $D_3$ is controlled by

$$|z_1 - z_2|^q \iint\limits_{|z_1 - z_2| \le r < \infty} \frac{r\,dr\,d\theta}{r^{2q}} = c'|z_1 - z_2|^{2-q}.$$

This yields our Hölder estimate, since $(2 - q)/q = 1 - 2/p$. $\square$

THEOREM 7.4. *Fix $0 < k < 1$, $R > 0$, and $p = p(k) > 2$ as above. For $\mu \in L^\infty(k, R)$, there is a function $f$ on $\mathbf{C}$, normalized so that $f(z) = z + \mathcal{O}(1/z)$ at $\infty$, with distribution derivatives satisfying the Beltrami equation $f_{\bar{z}} = \mu f_z$, and such that $f_{\bar{z}}$ and $f_z - 1$ belong to $L^p$. Any such $f$ is unique. The solution $f$ is a homeomorphism of $\mathbf{C}$, which is analytic on any open set on which $\mu = 0$. If $\mu \in C^1$ and $\mu_z \in C^1$, then $f \in C^1$.*

*Proof.* The proof of existence is now easy. Define $g \in L^p$ by (7.2), and define

$$f(z) = z + T(\mu g + \mu).$$

Since $T$ is a convolution operator with kernel $1/z$ locally in $L^1$, $f$ is continuous. Moreover, $f$ is normalized at $\infty$, and

$$\begin{aligned} f_{\bar{z}} &= \mu g + \mu, \\ f_z &= 1 + S(\mu g + \mu) = 1 + g \end{aligned}$$

in the sense of distributions, so $f$ satisfies the Beltrami equation. The defining formula for $T$ shows that $f$ is analytic on any open set on which $\mu = 0$.

To prove the uniqueness, suppose $F$ is another solution of the Beltrami equation with $F_{\bar{z}}, F_z - 1 \in L^p$. Then $G = F_z - 1$ satisfies $G = S(F_{\bar{z}})$. (We use here the fact that if $h$ has distribution derivatives in $L^p$, then $h_z = S(h_{\bar{z}})$.) As before we obtain $G = (I - U_\mu)^{-1}S(\mu)$, so $G$ coincides with $g$, $F_z$ with $f_z$, and $F_{\bar{z}}$ with $f_{\bar{z}}$. Thus $F = f + c$, and the normalization at $\infty$ gives $c = 0$.

Now suppose $\mu$ is smooth. Following [A2], to obtain the smoothness of $f$ we consider the differentiated Beltrami equation

$$(f_z)_{\bar{z}} = \mu(f_z)_z + \mu_z f_z .$$

If $f$ were $C^1$ we could write $\varphi = \log f_z$ and the equation would become $\varphi_{\bar{z}} = \mu\varphi_z + \mu_z$. Using the Cauchy–Green formula as before, applied to $\varphi$ instead of $f$ and using $\varphi(\infty) = 0$, we obtain $\varphi = T(\varphi_{\bar{z}})$. Consequently $\varphi_z = S(\varphi_{\bar{z}}) = S(\mu\varphi_z) + S(\mu_z)$, and

$$(I - U_\mu)\varphi_z = S(\mu_z).$$

This is the same type of equation as we dealt with above.

We now wish to run this argument backwards. Define

$$\psi = (I - U_\mu)^{-1}S(\mu_z) \in L^p,$$

which satisfies

$$\psi = S(\mu\psi) + S(\mu_z).$$

Define

$$\varphi = T(\mu\psi + \mu_z).$$

Since $\mu\psi + \mu_z \in L^p$ is supported on $\Delta(0, R)$, the proof of Theorem 7.3 shows that $\varphi$ is Hölder continuous with exponent $1 - 2/p$. The distribution derivatives of $\varphi$ are

$$
\begin{aligned}
\varphi_{\bar{z}} &= \mu\psi + \mu_z, \\
\varphi_z &= S(\mu\psi + \mu_z) = \psi,
\end{aligned}
$$

which belong to $L^p$. Thus if $h = e^\varphi$, we are justified in calculating the distribution derivatives of $h$ to be $h_z = e^\varphi\varphi_z$ and $h_{\bar{z}} = e^\varphi\varphi_{\bar{z}} = e^\varphi(\mu\varphi_z + \mu_z) = h_z\mu + h\mu_z = (h\mu)_z$. Hence $h\,dz + (h\mu)d\bar{z}$ is exact, and

$$f(z) = \int_0^z h\,dz + (h\mu)d\bar{z}$$

is well-defined. Since $\varphi$ (hence $h$) is continuous and $\mu$ is $C^1$, $f$ is $C^1$. Moreover, we have $f_z = h$ and $f_{\bar{z}} = h\mu$, so $f_{\bar{z}} = \mu f_z$ as desired. For $|z| > R$, $\mu h = 0$ so that $f$ is analytic there. Since $\varphi(\infty) = 0$, $h(z) = 1 + \mathcal{O}(1/z)$ at $\infty$, and $f(z) - z$ is analytic at $\infty$. Adding a constant to $f$, we can arrange that $f(z) = z + \mathcal{O}(1/z)$ at $\infty$. Also

$$J_f = (1 - |\mu|^2)|h|^2 = (1 - |\mu|^2)|e^{2\varphi}| \neq 0.$$

Thus $f$ is locally one-to-one, and $f$ is one-to-one near $\infty$, so $f$ is globally one-to-one, and $f$ is a homeomorphism of the Riemann sphere to itself. Thus $f \in QC^1(k, R)$.

To complete the proof it suffices to show that $f$ is a homeomorphism, even when $\mu$ is not smooth. For this we approximate $\mu$ pointwise a.e. by smooth Beltrami coefficients $\mu_n \in L^\infty(k, R)$, with corresponding $f_n \in QC^1(k, R)$. On account of the uniform Hölder estimates of Theorem 7.3, the sequence $f_n$ is equicontinuous, and we may assume $f_n$ converges uniformly to $F$ on $\overline{\mathbf{C}}$. By Lemma 7.1 and Theorem 7.3 the inverses $f_n^{-1}$ are also equicontinuous, so $f_n^{-1}$ converges uniformly to an inverse for $F$, and $F$ is a homeomorphism.

Let $g_n = (I - U_{\mu_n})^{-1}(S(\mu_n))$ as before. Choose $s > p$ so that $kK_s < 1$. Everything we have done works in $L^s$ as well as in $L^p$, and in particular $(I - U_{\mu_n})^{-1}$ is defined and bounded on $L^s$, so that $g_n$ is bounded in $L^s$. Since $\mu_n \to \mu$ a.e. (and since $\mu_n \in L^\infty(k, R)$), $\mu_n$ converges in $L^r$ for all $r < \infty$, and also $\mu_n g_n$ converges in $L^r$ for all $r < s$. Thus $g_n = S(\mu_n g_n) + S(\mu_n)$ converges in $L^p$, say to $g$. Since $f_n \to F$ uniformly, and $(f_n)_z = g_n + 1$, we see that $F_z = g + 1$ in the sense of distributions. Also $(f_n)_{\bar{z}} = \mu_n(f_n)_z$ converges to $\mu g$ in $L^p$, so $F_{\bar{z}} = \mu g$ in the sense of distributions. Thus $F$ coincides with the unique solution $f$ of the Beltrami equation with $L^p$ derivatives, and we are done. $\square$

We let $QC(k, R)$ denote the family of homeomorphisms $f$ of $\overline{\mathbf{C}}$ that are normalized solutions of the Beltrami equation with Beltrami coefficient $\mu \in L^\infty(k, R)$ as in Theorem 7.4. The proof of Theorem 7.4 shows the following.

THEOREM 7.5. *Suppose* $f_n \in QC(k, R)$ *has Beltrami coefficient* $\mu_n$, *and* $f \in QC(k, R)$ *has Beltrami coefficient* $\mu$. *If* $\mu_n \to \mu$ *a.e., then* $f_n \to f$ *uniformly, and* $(f_n)_z - 1$ *and* $(f_n)_{\bar{z}}$ *converge respectively to* $f_z - 1$ *and to* $f_{\bar{z}}$ *in* $L^p$ *for* $p = p(k)$ *as above. Furthermore the inverse functions* $f_n^{-1}$ *converge uniformly to* $f^{-1}$.

Theorem 7.5 shows that $f$ depends continuously on any continuous parameters. We also have analytic dependence of $f$ on analytic parameters.

THEOREM 7.6. *Suppose that* $\mu(z, t) \in L^\infty(k, R)$ *depends analytically on one or several parameters* $t$, *for each fixed* $z \in \mathbf{C}$. *Then the corresponding* $f(z, t)$ *also depends analytically on* $t$.

*Proof.* First note that if $\mu(z, t)$ depends analytically on $t$ for each fixed $z$, then on account of the uniform bounds on $\mu$ and the uniform estimates provided by the Schwarz inequality, the functions move analytically in the Banach space $L^\infty$ with the parameter $t$. Since the supports of $\mu(z, t)$ are uniformly bounded, the functions $\mu(z, t)$ also move analytically in $L^p$ with $t$. Thus $g = (I - U_\mu)^{-1}S(\mu)$ moves analytically in $L^p$ with $t$, and $T(\mu g + \mu)$ moves analytically with $t$, in the norm of uniform convergence on $\mathbf{C}$. Hence $f(z,t) = z + T(\mu g + \mu)$ depends analytically on $t$. $\square$

We complete this chapter with some remarks.

The uniform estimates derived in Theorems 7.2 and 7.3 clearly apply to $f \in QC(k, R)$ as do the computation rules for quasiconformal mappings and also Lemma 7.1. Also, we can assign to any $\mu \in L^\infty(k, R)$ a measurable ellipse field, and the arguments involving ellipse fields carry over to $f \in QC(k, R)$, as can be seen by approximating $\mu$ by smooth Beltrami coefficients.

We have normalized our quasiconformal maps so they look like $z + \mathcal{O}(1/z)$ near infinity. Given $\mu$, the corresponding $f$ is unique up to compositions with Möbius transformations, so instead of this normalization we could normalize by fixing any three points of the sphere.

We do not have to restrict ourselves to functions which are holomorphic in $\{|z| > R\}$. By composing with a Möbius transformation, it will always be enough to simply assume $f$ is holomorphic in some disk. Although this restriction can also be removed (see [A2]), in our applications this condition will always be satisfied.

The way we will use quasiconformal mappings to study dynamical systems is as follows. Suppose $g$ is a smooth function with certain dynamical behavior, and suppose we can construct an ellipse field $E$ that is invariant under $g$, corresponding to a measurable Beltrami coefficient $\mu$ with $|\mu| \leq k < 1$. Let $\varphi$ solve the corresponding Beltrami equation, and set $f = \varphi \circ g \circ \varphi^{-1}$. Then $f$ has the same dynamical behavior as $g$, and moreover $f$ is analytic. Indeed, $\varphi^{-1}$ maps infinitesimal circles to $E$, $g$ maps $E$ to $E$, and $\varphi$ then maps $E$ back to infinitesimal circles, so that $f$ maps infinitesimal circles to infinitesimal circles, and $f$ is analytic.

# II

# Fixed Points and Conjugations

The study of complex dynamical systems begins with the description of the local behavior near fixed points. We are concerned with the existence of canonical coordinate systems at fixed points. The coordinatizing functions play an important role, both locally and globally.

## 1. Classification of Fixed Points

Suppose $z_0$ is a fixed point of an analytic function $f$, that is, $f(z_0) = z_0$. The number $\lambda = f'(z_0)$ is called the *multiplier* of $f$ at $z_0$. We classify the fixed point according to $\lambda$ as follows:

*Attracting* : $|\lambda| < 1$. (If $\lambda = 0$ we refer to a *superattracting* fixed point.)

*Repelling* : $|\lambda| > 1$.

*Rationally neutral* : $|\lambda| = 1$ and $\lambda^n = 1$ for some integer $n$.

*Irrationally neutral* : $|\lambda| = 1$ but $\lambda^n$ is never 1.

It will be convenient to denote the iterates of a function $f$ by $f^1 = f$ and $f^n = f^{n-1} \circ f$.

Suppose $z_0$ is an attracting fixed point for $f$. If $|\lambda| < \rho < 1$, then $|f(z)-z_0| \le \rho|z-z_0|$ on some neighborhood of $z_0$. Then $|f^n(z)-z_0| \le \rho^n|z - z_0|$, and consequently the iterates $f^n$ converge uniformly to $z_0$ on the neighborhood. We define the *basin of attraction* of $z_0$, denoted by $A(z_0)$, to consist of all $z$ such that $f^n(z)$ is defined for all $n \ge 1$ and $f^n(z) \to z_0$. Thus for $\varepsilon > 0$ small, $A(z_0)$ coincides with the union of the backward iterates $f^{-n}(\Delta(z_0,\varepsilon))$, and consequently $A(z_0)$ is open. The connected component of $A(z_0)$ containing $z_0$ is called the *immediate basin of attraction* of $z_0$ and denoted by $A^*(z_0)$.

We say that a function $f : U \to U$ is *(conformally) conjugate* to a function $g : V \to V$ if there is a conformal map $\varphi : U \to V$ such that $g = \varphi \circ f \circ \varphi^{-1}$, that is, such that

$$\varphi(f(z)) = g(\varphi(z)). \tag{1.1}$$

The maps $f$ and $g$ can be regarded as the same map viewed in different coordinate systems. The definition implies the iterates $f^n$ and $g^n$ are also conjugate, $g^n = \varphi \circ f^n \circ \varphi^{-1}$, as are $f^{-1}$ and $g^{-1}$ when defined, $g^{-1} = \varphi \circ f^{-1} \circ \varphi^{-1}$. Note that $\varphi$ maps fixed points of $f$ to fixed points of $g$, and the multipliers at the corresponding fixed points are equal. A basin of attraction for $f$ is mapped by a conjugating map $\varphi$ onto a basin of attraction for $g$.

Suppose we are given

$$f(z) = z_0 + \lambda(z - z_0) + a(z - z_0)^p + \cdots$$

with $p \ge 2$. Near $z_0$ the function $f$ "looks like" $g(\zeta) = \lambda\zeta$ in the case $\lambda \ne 0$, and like $g(\zeta) = a\zeta^p$ in the case $\lambda = 0$, $a \ne 0$, where $\zeta = z - z_0$. Does there always exist a $\varphi$ conjugating $f$ to $g$? It turns out that the answer depends on $f$ and in particular on the multiplier $\lambda$. We will show that a conjugation exists in the cases of attracting and repelling fixed points. In the case of an irrationally neutral fixed point, a conjugation exists unless $\lambda$ is "very close" to roots of unity. In the case of a rationally neutral fixed point, a conjugation does not exist in general, but a conjugation to another canonical form exists in a large domain with the point as a cusp on its boundary.

If $\lambda \ne 0$ and $\lambda$ is not a root of unity, then the conjugation $\varphi$ is unique up to a scale factor. To prove this, it suffices to show that any conjugation of $f(z) = \lambda z$ to itself is a constant multiple of $z$. Suppose $\varphi(z) = a_1 z + a_2 z^2 + \cdots$ is such a conjugation, so that $\varphi(\lambda z) = \lambda\varphi(z)$. Substituting power series and equating coefficients, we obtain $a_n \lambda^n = \lambda a_n$, so that $a_n = 0$ for $n \geq 2$, and $\varphi(z) = a_1 z$. In the superattracting case $\lambda = 0$, the same method shows that any conjugation to $\zeta^p$ is unique up to multiplication by a $(p-1)$th root of unity. The functional equation of a conjugation of $z^p$ to itself is $\varphi(z^p) = \varphi(z)^p$, and comparing power series we find that $\varphi(z) = a_1 z$, where $a_1^p = a_1$.

When $\lambda$ is an $n$th root of unity, any conformal map of the form $\varphi(z) = zh(z^n) = b_1 z + b_{n+1} z^{n+1} + \cdots$ conjugates $\lambda z$ to itself. In particular, the identity map $z$ with multiplier $\lambda = 1$ is conjugated to itself by any conformal $\varphi$. We must consider other normal forms.

The idea of conformal conjugation was introduced by E. Schröder (1871) to study iteration of rational functions. Equation (1.1) and its variants are referred to as *Schröder's equation*. There was some continuing interest in the iteration of rational functions in connection with algorithms for approximating roots, specifically Newton's method. Schröder was interested in finding effective methods for computing iterates. Schröder (1870), and also A. Cayley (1879), found the basins of attraction for Newton's method applied to quadratic polynomials (see the example to follow), and both mention the cubic case as an interesting problem. (According to Douady (1986), it was a question from a student in 1978 about the convergence of Newton's method that initially aroused the interest of J. Hubbard, and "by contamination" of Douady himself, in rational iteration theory.) For color pictures of domains of attraction associated with Newton's method applied to certain cubic polynomials, see Maps 66, 77 and 78 on pages 91, 116 and 117 of [**PeR**].

Schröder emphasized how addition and multiplication formulae for trigonometric and elliptic functions give rise to conjugations. For instance, the double angle formula for tangent shows that $z = -i \tan \zeta$ conjugates $2z/(1+z^2)$ to $2\zeta$. The double angle formula for cosine is behind the following example.

EXAMPLE. The polynomial $P(z) = z^2 - 2$ has a superattracting fixed point at $\infty$. Consider the conformal map $h(\zeta) = \zeta + 1/\zeta$ of $\{|\zeta| > 1\}$ onto $\mathbf{C}\backslash[-2,2]$. The identity $P(h(\zeta)) = h(\zeta)^2 - 2 = h(\zeta^2)$ gives $h^{-1} \circ P \circ h = \zeta^2$, and $P(z)$ is conjugate to $\zeta^2$. Thus the dynamics of $P(z)$ on $\mathbf{C}\backslash[-2,2]$ are the same as those of $\zeta^2$ on $\{|\zeta| > 1\}$. Since the iterates of any $\zeta$, $|\zeta| > 1$, under $\zeta^2$ tend to $\infty$, so do the iterates under $P$ of any $z \in \mathbf{C}\backslash[-2,2]$. Evidently $[-2,2]$ is invariant under $P$, so the basin of attraction of $\infty$ for $P$ is $A(\infty) = \overline{\mathbf{C}}\backslash[-2,2]$.

EXAMPLE. The preceding example extends as follows. Let $T_n(z) = \cos(n \arccos z)$ be the $n$th Tchebycheff polynomial, and set $F_n(z) = 2T_n(z/2)$, which is monic. The coordinate change $z = h(\zeta) = \zeta + 1/\zeta = e^{iw} + e^{-iw} = 2\cos w$ yields

$$F_n(h(\zeta)) = 2\cos(n \arccos(z/2)) = 2\cos(nw) = h(\zeta^n).$$

Thus $h$ implements a conjugation of $F_n$ and $\zeta^n$.

EXAMPLE. Let $S$ denote the unit square in the $\zeta$-plane with corners $0, 1, 1+i, i$ and let $h$ be the conformal map taking $S$ to the upper half-plane, with $0, 1, 1+i$ being mapped to $\infty, -1, 0$ respectively. Symmetry considerations show $h$ maps $i$ to $+1$ and $(1+i)/2$ to $i$. We continue $h$ by reflection to a meromorphic function on the whole $\zeta$-plane mapping unit-squares alternately to upper and lower half-planes. Then $h$ is doubly periodic, with periods 2 and $2i$ and with a double pole at each period point. The function $h(2\zeta)$ is also doubly periodic, and by comparing poles and expansion at 0 one checks that

$$h(2\zeta) = \frac{1}{4} \frac{\prod[h(\zeta) - h((\pm 1 \pm i)/2)]}{h(\zeta)(h(\zeta)-1)(h(\zeta)+1)} = \frac{(h(\zeta)^2+1)^2}{4h(\zeta)(h(\zeta)^2-1)}.$$

Thus the rational function $f(z) = (z^2+1)^2/4z(z^2-1)$ satisfies the Schröder equation $h(2\zeta) = f(h(\zeta))$, and $h$ implements a conjugation of $f(z)$ and multiplication by 2, at least where $h$ is univalent. In any event this makes the dynamic behavior of $f$ transparent. In particular, one sees that a point $z$ is iterated by $f$ to the repelling fixed point of $f$ at $\infty$ if and only if $z = h(\zeta)$ for some dyadic point $\zeta = 2^{-j}m + 2^{-k}ni$. Note that $h$ is not univalent at 0, and the multiplier of $f$ at $\infty$ is 4. The function $h$ is the Weierstrass $\mathcal{P}$-function. Several other elliptic functions were treated by Schröder, and this specific example was given by S. Lattès (1918).

EXAMPLE. We apply Newton's method (actually due in this form in 1690 to J. Raphson) to a quadratic polynomial $P(z)$ with simple zeros. We are iterating $f(z) = z - P(z)/P'(z)$, which has superattracting fixed points at the zeros of $P(z)$ and a repelling fixed point at $\infty$. The Möbius transformation $\zeta = \varphi(z)$ sending the two zeros of $P(z)$ to 0 and $\infty$ respectively, and $\infty$ to 1, conjugates $f(z)$ to $\zeta^2$. Since the midpoint of the line segment joining the zeros of $P(z)$ is mapped by $f$ to $\infty$, the midpoint is sent by $\varphi$ to the preimage $-1$ of

+1. Thus the perpendicular bisector of the line segment joining the zeros of $P(z)$ is mapped by $\varphi$ to the unit circle. We conclude that the basins of attraction for Newton's method are the respective open half-planes on either side of the bisector.

## 2.  Attracting Fixed Points

The easiest to treat are the attracting fixed points that are not super-attracting. The following linearization theorem is due to G. Koenigs (1884).

THEOREM 2.1. *Suppose $f$ has an attracting fixed point at $z_0$, with multiplier $\lambda$ satisfying $0 < |\lambda| < 1$. Then there is a conformal map $\zeta = \varphi(z)$ of a neighborhood of $z_0$ onto a neighborhood of $0$ which conjugates $f(z)$ to the linear function $g(\zeta) = \lambda\zeta$. The conjugating function is unique, up to multiplication by a nonzero scale factor.*

*Proof.* Suppose $z_0 = 0$. Define $\varphi_n(z) = \lambda^{-n} f^n(z) = z + \cdots$. Then $\varphi_n$ satisfies

$$\varphi_n \circ f = \lambda^{-n} f^{n+1} = \lambda\varphi_{n+1}.$$

Thus if $\varphi_n \to \varphi$, then $\varphi \circ f = \lambda\varphi$, so $\varphi \circ f \circ \varphi^{-1} = \lambda\zeta$, and $\varphi$ is a conjugation.

To show convergence note that for $\delta > 0$ small

$$|f(z) - \lambda z| \le C|z|^2, \qquad |z| \le \delta.$$

Thus $|f(z)| \le |\lambda||z| + C|z|^2 \le (|\lambda| + C\delta)|z|$, and by induction with $|\lambda| + C\delta < 1$,

$$|f^n(z)| \le (|\lambda| + C\delta)^n |z|, \qquad |z| \le \delta.$$

We choose $\delta > 0$ so small that $\rho = (|\lambda| + C\delta)^2/|\lambda| < 1$, and we obtain

$$|\varphi_{n+1}(z) - \varphi_n(z)| = \left| \frac{f^n(f(z)) - \lambda f^n(z)}{\lambda^{n+1}} \right| \le \frac{C|f^n(z)|^2}{|\lambda|^{n+1}} \le \frac{\rho^n C|z|^2}{|\lambda|}$$

for $|z| \le \delta$. Hence $\varphi_n(z)$ converges uniformly for $|z| \le \delta$, and the conjugation exists. The uniqueness assertion was already noted in the preceding section. □

The conjugation $\varphi$ constructed above is normalized so that $\varphi'(z_0) = 1$. The uniform convergence shows that if $f$ depends analytically on a parameter, so does the (normalized) conjugation $\varphi$.

The Schröder equation (1.1) satisfied by the conjugating function $\varphi$ becomes

$$\varphi(f(z)) = \lambda\varphi(z). \tag{2.1}$$

This equation allows us to extend $\varphi$ analytically to the entire basin of attraction $A(z_0)$, by the formula $\varphi(z) = \varphi(f^n(z))/\lambda^n$, where $n$ is chosen large enough so that $f^n(z)$ belongs to a coordinate neighborhood of $z_0$. The extended $\varphi$ is well-defined and satisfies the same functional equation (2.1). Note that $\varphi(z) = 0$ if and only if $f^n(z) = z_0$ for some $n \geq 1$.

The branch of $\varphi^{-1}$ mapping $0$ to $z_0$ can be continued until we meet a critical point of $f$ or leave the domain of $f$. If $f$ is a polynomial, or rational, then the range of $\varphi$ covers the entire complex plane. The Riemann surface of $\varphi^{-1}$ is a branched covering surface over $\mathbf{C}$, on which the action of $f$ can be described as a multiplication by $\lambda$ and a shift from one sheet to another.

EXAMPLE. Suppose $f(z)$ is a finite Blaschke product of order $d$, with a simple zero at $z = 0$:

$$f(z) = e^{i\theta_0} z \prod_{j=2}^{d} \frac{z - a_j}{1 - \overline{a_j} z}.$$

The basin of attraction $A(0)$ of the fixed point $0$ is the open unit disk $\Delta$. The conjugating function $\varphi$ is an infinite-to-one mapping of $\Delta$ onto the complex plane $\mathbf{C}$, whose zero set accumulates on all of $\partial\Delta$. The functional equation (2.1) for $\varphi$ shows that the critical points of $\varphi$ are the critical points of $f$ and all their inverse iterates.

## 3. Repelling Fixed Points

The existence of a conjugating map for a repelling fixed point follows immediately from the attracting case. For suppose $f(z) = z_0 + \lambda(z - z_0) + \cdots$ where $|\lambda| > 1$. Then $f^{-1}(z) = z_0 + (z - z_0)/\lambda + \cdots$ has an attracting fixed point at $z_0$. Any map conjugating $f^{-1}(z)$ to $\zeta/\lambda$ also conjugates $f(z)$ to $\lambda\zeta$.

EXAMPLE. If $P(z)$ is a polynomial with a repelling fixed point at the origin, the corresponding map $h = \varphi^{-1}$ can be defined on the entire plane by the functional equation $h(\lambda\zeta) = P(h(\zeta))$. It is an entire function of finite order, and in fact it satisfies an estimate of the form

$$\log|h(\zeta)| \le C_0 + C_1|\zeta|^\tau, \qquad \tau = \frac{\log d}{\log|\lambda|}, \quad d = \deg P.$$

As a special case, the monomial $z^m$ has a repelling fixed point at $z = 1$ with multiplier $m$. The coordinate change $\zeta = \varphi(z) = \log z$ conjugates $z^m$ to $m\zeta$. The corresponding entire function is $h(\zeta) = e^\zeta$.

## 4.   Superattracting Fixed Points

In the superattracting case the existence of a conjugation was first proved by L.E. Boettcher (1904).

THEOREM 4.1. *Suppose $f$ has a superattracting fixed point at $z_0$,*

$$f(z) = z_0 + a_p(z - z_0)^p + \cdots, \qquad a_p \ne 0, \, p \ge 2.$$

*Then there is a conformal map $\zeta = \varphi(z)$ of a neighborhood of $z_0$ onto a neighborhood of $0$ which conjugates $f(z)$ to $\zeta^p$. The conjugating function is unique, up to multiplication by a $(p-1)$th root of unity.*

*Proof.* Suppose $z_0 = 0$. For $|z|$ small there is $C > 1$ such that $|f(z)| \le C|z|^p$. By induction, writing $f^{n+1} = f^n \circ f$ and using $p \ge 2$, we find that

$$|f^n(z)| \le (C|z|)^{p^n}, \qquad |z| \le \delta,$$

so $f^n(z) \to 0$ super-exponentially.

If we change variables by setting $w = cz$ where $c^{p-1} = 1/a_p$ then we have conjugated $f$ to the form $f(w) = w^p + \cdots$. Therefore we may assume $a_p = 1$. We wish to find a conjugating map $\varphi(z) = z + \cdots$ such that $\varphi(f(z)) = \varphi(z)^p$, which is equivalent to the condition that $\varphi \circ f \circ \varphi^{-1} = \zeta^p$. Let

$$\varphi_n(z) = f^n(z)^{p^{-n}} = (z^{p^n} + \cdots)^{p^{-n}} = z(1 + \cdots)^{p^{-n}},$$

which is well defined in a neighborhood of the origin. The $\varphi_n$'s satisfy

$$\varphi_{n-1} \circ f = (f^{n-1} \circ f)^{p^{-n+1}} = \varphi_n^p,$$

so if $\varphi_n \to \varphi$ then $\varphi$ satisfies $\varphi \circ f = \varphi^p$ and so is a solution. To show that $\{\varphi_n\}$ converges, we write $f^{n+1} = f \circ f^n$ and note that

$$
\frac{\varphi_{n+1}}{\varphi_n} = \left(\frac{\varphi_1 \circ f^n}{f^n}\right)^{p^{-n}} = (1 + \mathcal{O}(|f^n|))^{p^{-n}}
$$
$$
= 1 + \mathcal{O}(p^{-n})\mathcal{O}(|z|^{p^n} C^{p^n}) = 1 + \mathcal{O}(p^{-n})
$$

if $|z| \leq 1/C$. Thus the product

$$
\prod_{n=1}^{\infty} \frac{\varphi_{n+1}}{\varphi_n}
$$

converges uniformly for $|z| \leq c < 1/C$, and this implies $\{\varphi_n\}$ converges. Hence $\varphi$ exists. The uniqueness statement was noted in Section 1. $\square$

Again if $f$ depends analytically on a parameter, then so does the conjugation $\varphi$ constructed above. In this case the functional equation satisfied by the Boettcher coordinate function,

$$
\varphi(f(z)) = \varphi(z)^p,
$$

allows us to extend $\varphi$ analytically only until we meet a critical point of $f$. However, the functional equation

$$
\log|\varphi(f(z))| = p\log|\varphi(z)|
$$

allows us to extend $\log|\varphi(z)|$ to the entire basin of attraction $A(z_0)$ of $z_0$. The extended function is a negative harmonic function, except for logarithmic poles at all inverse iterates of $z_0$.

Consider the case in which $f$ is a $p$-sheeted cover of the immediate basin of attraction $A^*(z_0)$ onto itself. The only pole of $\log|\varphi(z)|$ is now $z_0$, and $\log|\varphi(z)| = \log|z - z_0| + \mathcal{O}(1)$ at $z_0$. Furthermore, the functional equation shows that $\log|\varphi(z)| \to 0$ as $z \to \partial A^*(z_0)$. Except for sign, these are precisely the properties that characterize Green's function $G(z, z_0)$ for $A^*(z_0)$ with pole at $z_0$, and we obtain

$$
\log|\varphi(z)| = -G(z, z_0), \qquad z \in A^*(z_0).
$$

EXAMPLE. A polynomial $P(z) = az^d + \cdots$ with $d \geq 2$ and $a \neq 0$ has a superattracting fixed point at $\infty$. Replacing 0 by $\infty$ in Boettcher's theorem, we see that $P(z)$ is conjugate to $\zeta^d$ near $\infty$. The conjugating map has the form $\varphi(z) = cz + \mathcal{O}(1)$, with a simple pole at $\infty$. Again the functional equation

$$\log|\varphi(P(z))| = d\log|\varphi(z)|, \qquad |z| > R,$$

allows us to extend $\log|\varphi|$ to the entire basin of attraction $A(\infty)$ of $\infty$. The basin of attraction has no bounded components, by the maximum principle, so that $A(\infty) = A^*(\infty)$ is connected. The extended $\log|\varphi|$ is harmonic on $A(\infty)$, and $\log|\varphi(z)| \to 0$ as $z$ tends to the boundary. This time

$$\log|\varphi(z)| = \log|z| + \log|c| + o(1), \qquad |z| \to \infty.$$

Now Green's function for an exterior domain $D$ with pole at $\infty$ is characterized as the positive harmonic function on $D$ which tends to zero at $\partial D$ (or at least at all regular boundary points of $\partial D$), and which has the form

$$G(z, \infty) = \log|z| + \sigma + o(1)$$

at $\infty$. The constant $\sigma$ appearing here is called *Robin's constant*, and the logarithmic capacity of $\partial D$ can be defined as $e^{-\sigma}$. (See [Ts].) In the case at hand we see that $\log|\varphi(z)|$ is Green's function for $A(\infty)$ with pole at $\infty$, and $\log|c|$ is Robin's constant. The logarithmic capacity of $\partial A(\infty)$ is $1/|c| = |a|^{-1/(d-1)}$.

# 5.   Rationally Neutral Fixed Points

Suppose $\lambda^n = 1$ and $f(z) = \lambda z + az^{p+1} + \cdots$, $a \neq 0$. We consider three cases:

1. $\lambda = 1, p = 1$,

2. $\lambda = 1, p > 1$,

3. $\lambda^n = 1, \lambda \neq 1$.

CASE 1. We assume $f(z) = z + az^2 + \cdots$ with $a \neq 0$. By conjugating $f$ by $\varphi(z) = az$, we may assume $a = 1$. Let us next move 0 to $\infty$

by conjugating by the inversion $z \rightarrow -1/z$. The conjugated map has the expansion

$$g(z) = z + 1 + \frac{b}{z} + \cdots \qquad (5.1)$$

near $\infty$. We shall consider two methods for proving there is a map $\varphi$ conjugating $g$ to the translation $z \rightarrow z + 1$. The first is due to Fatou and uses the asymptotic behavior of $g^n$. The second involves quasiconformal mappings.

First observe that if $C_0 > 0$ is sufficiently large, then the half-plane $\{\mathrm{Re}\, z > C_0\}$ is invariant under $g$, and

$$\mathrm{Re}\, g^n(z) > \mathrm{Re}\, z + \frac{n}{2}, \qquad \mathrm{Re}\, z \geq C_0,\, n \geq 1.$$

This is proved easily by induction, as is the upper bound in

$$\frac{n}{2} \leq |g^n(z)| \leq |z| + 2n, \qquad \mathrm{Re}\, z \geq C_0,\, n \geq 1.$$

These estimates are valid whenever $g(z) = z + 1 + o(1)$, and they will be used later.

In the first method we define $\varphi_n$ by

$$\varphi_n(z) = g^n(z) - n - b \log n, \qquad \mathrm{Re}\, z \geq C_0.$$

From

$$g^{k+1}(z) = g^k(z) + 1 + \frac{b}{g^k(z)} + \mathcal{O}\left(\frac{1}{k^2}\right)$$

we obtain

$$\varphi_{k+1}(z) - \varphi_k(z) = b[\log k - \log(k+1)] + \frac{b}{g^k(z)} + \mathcal{O}\left(\frac{1}{k^2}\right) = \mathcal{O}\left(\frac{1}{k}\right),$$

where the estimates are all independent of $z$. It follows that

$$|\varphi_n(z) - z| \leq |\varphi_1(z) - z| + \sum_{k=1}^{n-1} |\varphi_{k+1}(z) - \varphi_k(z)| = \mathcal{O}(\log n)$$

for $\mathrm{Re}\, z \geq C_0$. Now to prove that $\varphi_n(z)$ converges, we estimate

$$\begin{aligned}
\varphi_{n+1}(z) - \varphi_n(z) &= b \log n - b \log(n+1) + g^{n+1}(z) - g^n(z) - 1 \\
&= -\frac{b}{n} + \frac{b}{g^n(z)} + \mathcal{O}\left(\frac{1}{n^2}\right) \\
&= b\left[\frac{1}{n + b \log n + \varphi_n(z)} - \frac{1}{n}\right] + \mathcal{O}\left(\frac{1}{n^2}\right) \\
&= \frac{b}{n^2}\mathcal{O}\left(|b \log n + \varphi_n(z)|\right) + \mathcal{O}\left(\frac{1}{n^2}\right) = \mathcal{O}\left(\frac{\log n}{n^2}\right).
\end{aligned}$$

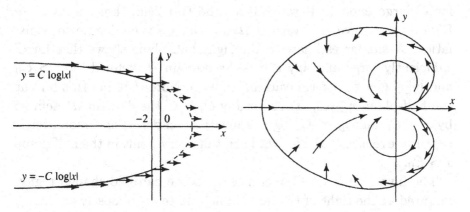

FIGURE 1. Parabolic attracting regions at $\infty$ and at $0$.

Hence $\sum |\varphi_{n+1}(z) - \varphi_n(z)| < \infty$, and $\varphi_n$ converges, say to $\varphi$. Since the $\varphi_n$'s are univalent, so is $\varphi$. From the definition, $\varphi_n(g(z)) = \varphi_{n+1}(z) + 1 + b\log(1 + 1/n)$. Hence $\varphi \circ g = \varphi + 1$, and $\varphi$ conjugates $g$ to the translation $z \to z + 1$.

Using the functional equation $\varphi(g(z)) = \varphi(z) + 1$, we can continue $\varphi$ analytically to any domain $\Omega$ on which $g$ is defined and satisfies $g(\Omega) \subseteq \Omega$ and $\operatorname{Re} g^n(z) \to +\infty$, $z \in \Omega$. For instance, from the asymptotic form of $g$ at $\infty$ we see that for any $\delta > 0$ we can choose $C_\delta$ so large that $\Omega = \{|y| > -\delta x + C_\delta\} \cup \{x > C_0\}$, the union of three half-spaces, is invariant under $g$, and $g$ on $\Omega$ is conjugate to translation.

With some more effort we can construct an invariant domain $\Omega$ with smooth boundary so that the part of $\Omega$ in the half-plane $x < -2$ consists of the domain above the curve $y = C\log|x|$ and the symmetric domain below the curve $y = -C\log|x|$, where $C > 0$ is appropriately large, as at the left of Figure 1. Indeed, from $g(z) = u + iv = x + 1 + iy + \mathcal{O}(1/|z|)$, we obtain for $z$ on the upper boundary curve that

$$\log|u| = \log|x| - \frac{1}{|x|} + \mathcal{O}\left(\frac{1}{|x|^2}\right),$$

so that

$$
\begin{aligned}
v &\geq y - \frac{A}{|z|} = C\log|x| - \frac{A}{|z|} \\
&\geq C\log|u| + \frac{C}{|x|} - \frac{A}{|z|} + \mathcal{O}\left(\frac{1}{|x|^2}\right) \geq C\log|u|
\end{aligned}
$$

for $C$ large enough. If we connect the two logarithm curves as in Figure 1, we have an invariant $\Omega$, on which $g$ is conjugate to translation. A similar estimate in the right half-plane shows that for $C$ sufficiently large and any $t \geq 0$ the domain $\Omega_t$ defined by $x > C_0$ and $|y| \leq C \log x + t$ is invariant under $g$. Note that the $\Omega_t$'s fill out the half-plane $x \geq C_0$. Thus for any $C_1 > C$ the domain $\Omega^*$ defined by $x > C_0$ and $|y| \leq C_1 \log x$ is invariant under $g$, and moreover it captures eventually the forward orbit of every point in the half-plane $x > C_0$.

The inversion $z \to -1/z$ carries $\Omega$ to the cardioid-shaped region pictured at the right of Figure 1, which we refer to loosely as the *attracting petal* for the fixed point. Inside the attracting petal $f$ is conjugate to $z \to z/(1-z)$. At the cusp the boundary curves are tangent to the positive real axis. In fact, the images of the logarithm curves defined above have the form $y = \pm C x^2 \log(1/x) + \mathcal{O}(y^2)$ as $x \to 0+$. The iterates of any point in the cardioid tend to 0 in a direction tangent to the negative real axis. Furthermore they all eventually enter and approach 0 through the image of $\Omega^*$, which forms a narrow corridor bounded by curves of the form $y = \pm C_1 x^2 \log(1/|x|) + \mathcal{O}(y^2)$ as $x \to 0-$.

We now describe the second method for dealing with this case. As before, assume $g$ is given by (5.1), and let $V = \{\mathrm{Re}\, z > A\}$ be a half-plane such that $g(V) \subset V$. Let $L$ be the vertical line bounding $V$. Then $L' = g(L)$ is a smooth curve which is approximately the translation of $L$ by 1 to the right. Let $S$ denote the vertical strip in the $z$-plane bounded by $L$ and $L'$. In the $\zeta = \xi + i\eta$ plane, let $\Sigma$ be the honest strip $\{0 < \xi < 1\}$. We construct a diffeomorphism $h : \Sigma \to S$ by setting $h(i\eta) = A + i\eta$ and $h(1 + i\eta) = g(A + i\eta)$ and then filling in these boundary values in a smooth way, uniformly as $\mathrm{Im}\, \zeta \to \pm\infty$. For technical reasons, let $h$ be conformal on some disk in $\Sigma$.

We now extend $h(\zeta)$ to a diffeomorphism of the right half-plane onto $V$ by iterating the formula $h(\zeta + 1) = g(h(\zeta))$. The analyticity of $g$ guarantees that the ellipse field (Beltrami coefficient) $\mu$ of $h$ is invariant under translation by 1, $\mu(\zeta + 1) = \mu(\zeta)$. Extend $\mu$ to the whole $\zeta$-plane to be periodic. Then $\mu$ is smooth except at $\infty$, and $|\mu| \leq k < 1$. Also, $\mu = 0$ on open sets where $h$ is analytic.

Solve the Beltrami equation $\psi_{\bar{\zeta}} = \mu \psi_\zeta$, so that $\psi$ fixes the points $0, 1, \infty$. The ellipse field corresponding to $\mu$ is invariant under the

translation function $G(\zeta) = \zeta + 1$. Hence $\psi_1 = \psi \circ G \circ \psi^{-1}$ is conformal everywhere, as is seen by considering its action on infinitesimal
circles. Thus it is Möbius, and since $\psi_1$ fixes $\infty$ and takes 0 to 1 it
must be of the form $\psi_1(z) = az + 1$. So $\psi$ must satisfy the functional
equation $\psi(\zeta + 1) = a\psi(\zeta) + 1$.

Define $\varphi = \psi \circ h^{-1}$ on $V$. Then $\varphi(z)$ is analytic on $V$ and satisfies

$$\varphi \circ g = \psi \circ h^{-1} \circ g = \psi \circ G \circ h^{-1} = a(\psi \circ h^{-1}) + 1 = a\varphi + 1.$$

We claim that $a = 1$, and for this we check the stretching of $\varphi$
at $\infty$. Note that $h^{-1}$ can be extended from $V$ to to a quasiconformal homeomorphism of $\overline{\mathbf{C}}$, so that $\varphi$ is the restriction to $V$ of a
quasiconformal homeomorphism of $\overline{\mathbf{C}}$. The Hölder condition of Theorem I.7.3, applied to $1/\varphi(1/z)$, then yields an estimate of the form
$|\varphi(g^n(z))| \leq C|g^n(z)|^\alpha$ for some fixed $0 < \alpha < 1$, which is $\sim n^\alpha$. On
the other hand, if $a \neq 1$ then

$$\varphi(g^n(z)) = a^n\varphi(z) + a^{n-1} + \cdots + a + 1 = a^n\varphi(z) + (a^n - 1)/(a - 1).$$

Since $\varphi(g^n(z)) \to \infty$, we have $|a| > 1$, and then $|\varphi(g^n(z))| \sim |a|^n$,
which grows faster that $n^\alpha$. We conclude that $a = 1$, and $\varphi$ is the
desired conjugation.

CASE 2. Suppose now that $z' = f(z) = z + az^{p+1} + \cdots$ with $a \neq 0$
and $p > 1$. As before we may assume $a = 1$. Define $z = \zeta^{1/p}$ and
$z' = \zeta'^{1/p}$ for $0 < \arg \zeta$, $\arg \zeta' < 2\pi$ and $z, z'$ restricted to an
appropriate sector of aperture $2\pi/p$. Then

$$\zeta' = \zeta + p\zeta^2 + \mathcal{O}(|\zeta|^{2+1/p}).$$

We renormalize again to remove the $p$ and then change variables by
$\zeta = -1/z$ and $\zeta' = -1/z'$. This gives

$$z' = g(z) = z + 1 + \mathcal{O}(|z|^{-1/p}).$$

The first method in Case 1 can be used here again, but it becomes
much more complicated since not only $\log n$ enters but also other
terms involving $n^{k/p}$ for $1 \leq k \leq p$. On the other hand, the second
method involving quasiconformal maps goes through in this case
without change. Moreover, the discussion of the invariant domain $\Omega$
remains intact, except that now the invariant $\Omega$ is bounded by the
curves $y = \pm C|x|^{1-1/p}$ in some half-space $x < -M$.

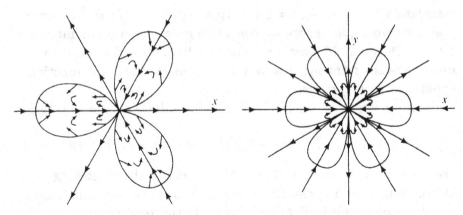

FIGURE 2. Pattern of attracting petals for $z + z^4$ and $-z + z^4$.

Another proof covering Case 2 proceeds as follows. First we kill off one by one the higher order Taylor coefficients of $f(z) = z + z^{p+1} + a_{p+2}z^{p+2} + \cdots$ by conjugations of the form $z \to z + \alpha z^k$. Only the coefficient of $z^{2p+1}$ refuses to die, and we obtain for any arbitrarily large prescribed $q$ a conjugation of $f(z)$ to

$$F(z) = z + z^{p+1} + Az^{2p+1} + \mathcal{O}(|z|^q).$$

The coefficient $A$ is actually a conformal conjugation invariant, though not the only one (see [**Vo**]). If now we fix $q > 2p + 1$, then the method used in Case 1 applied to $F(z)$ converges to the desired conjugation. (We mention as another approach the proof in [**Mi2**], in which a couple of preliminary smoothings near $\infty$ lead to rapid convergence.)

The picture is as follows. Instead of having a single attracting petal, there are now $p$ attracting petals, each within a sector of aperture $2\pi/p$, as in Figure 2. The angles of the boundary rays of the sectors are given in terms of the coefficient $a$ of $z^{p+1}$ by

$$\theta_k = -\frac{\arg a}{p} + \frac{2\pi k}{p}, \qquad 0 \le k \le p - 1.$$

These are called the *repelling directions*. They are exactly the angles for which $|1+az^p|$ is the largest, that is, for which $f(z) = z(1+az^p) + \mathcal{O}(|z^{p+2}|)$ expands the most. The gap between two consecutive petals is contained in a cusp bounded by curves with $|\theta - \theta_k| \sim |z|^{1/p}$. The directions for which $|1 + az^p|$ is smallest are given by $\theta'_k = \theta_k + \pi/p$, and these are called the *attracting directions*. The rays in these directions bisect the petals. Each petal is invariant under $f$, and the iterates of a point in the petal approach the fixed point 0 eventually through a corridor bounded by curves with $|\theta - \theta'_k| \sim |z|^{1/p}$. The symmetry of the conditions is easy to explain. Replacing $f(z) = z + az^{p+1} + \cdots$ by $\tilde{f}(z) = z - az^{p+1} - \cdots$ has the effect of running the flow backward, interchanging attracting and repelling directions.

CASE 3. Finally assume $f(z) = \lambda z + \cdots$, where $\lambda$ is a primitive $n$th root of unity. Then $f^n(z)$ must belong to either Case 1 or 2. Let $P_1, ..., P_p$ be the petals for $f^n$. Then $f(P_i)$ essentially coincides with $P_j$ for some $j$, and $f$ permutes the petals in cycles of length $n$. Thus $p = k \cdot n$ for some integer $k$, and $n$ divides $p$. In particular, if $f^n(z) = z + a_m z^m + \cdots$ where $a_m \neq 0$, then $m = kn + 1$ for some integer $k$, and the number of petals is $p = m - 1$. This is a purely algebraic fact, which is not so easy to prove directly.

EXAMPLE. If $f(z) = -z + z^4$, then $f^2(z) = z - 4z^7 + 6z^{10} - 6z^{13} + z^{16}$, so there are six petals and $f$ interchanges opposing petals. One can check that the perturbed map $-z + z^4 + \varepsilon z^5$ has fewer petals.

# 6.   Irrationally Neutral Fixed Points

Now let $\lambda = e^{2\pi i\theta}$ where $\theta$ is irrational. We want a solution $\varphi$ of the Schröder equation $\varphi(f(z)) = \lambda\varphi(z)$, normalized by $\varphi'(0) = 1$. For $h = \varphi^{-1}$ this becomes

$$f(h(\zeta)) = h(\lambda\zeta), \qquad h'(0) = 1. \tag{6.1}$$

We begin our discussion with a simple observation.

THEOREM 6.1. *A solution $h$ to the Schröder equation (6.1) in $\{|\zeta| < r\}$ is univalent.*

*Proof.* Suppose $h(\zeta_1) = h(\zeta_2)$. Then $h(\lambda^n \zeta_1) = h(\lambda^n \zeta_2)$ for all $n > 0$. But $\{\lambda^n\}$ is dense in the unit circle so $h(\zeta_1 e^{i\theta}) \equiv h(\zeta_2 e^{i\theta})$ for every $\theta$. This implies $h(\zeta_1 z) = h(\zeta_2 z)$ for $|z| < 1$, and since $h'(0) = 1$, $\zeta_1 = \zeta_2$. $\square$

THEOREM 6.2. *A solution $h$ to the Schröder equation (6.1) exists if and only if the sequence of iterates $\{f^n\}$ is uniformly bounded in some neighborhood of the origin.*

*Proof.* If $h$ exists then $f^n(z) = h(\lambda^n h^{-1}(z))$ is obviously bounded. On the other hand, if $|f^n| \leq M$ for all $n$, then we can define

$$\varphi_n(z) = \frac{1}{n} \sum_0^{n-1} \lambda^{-j} f^j(z).$$

Then $\{\varphi_n\}$ is a uniformly bounded sequence of analytic functions, so contains a convergent subsequence. Since $\varphi_n \circ f = \lambda \varphi_n + \mathcal{O}(1/n)$, any limit of the $\varphi_n$'s satisfies $\varphi \circ f = \lambda \varphi$. From $\lambda = f'(0)$ we have $\varphi_n'(0) = 1$, and $\varphi'(0) = 1$. Thus $h = \varphi^{-1}$ is a solution to (6.1). $\square$

One immediate consequence of Theorem 6.2, which is not so easy to see directly, is that if $f$ is topologically conjugate to $\lambda z$ near 0, then it is conformally conjugate. Indeed, if there is a topological mapping $h$, $h(0) = 0$, such that $h^{-1} \circ f \circ h = \lambda \zeta$, then for $\delta > 0$ small the images of the disks $\Delta(0, \delta)$ under $h$ are invariant under $f$, so the iterates of $f$ are uniformly bounded on a neighborhood of 0.

**THEOREM 6.3.** *There exists a $\lambda = e^{2\pi i \theta}$ so that the Schröder equation (6.1) has no solution for any polynomial $f$.*

*Proof.* Let $f(z) = z^d + \cdots + \lambda z$, and suppose there is a conjugation $h$ defined on $\Delta(0, \delta)$. Consider the $d^n$ fixed points of $f^n$, that is, the roots of

$$f^n(z) - z = z^{d^n} + \cdots + (\lambda^n - 1)z = 0.$$

One root is 0. Label the others $z_1, ..., z_{d^n - 1}$ and note that since $f^n(z) = h(\lambda^n h^{-1}(z))$ has only one zero in $\Delta(0, \delta)$, $z_j \notin \Delta(0, \delta)$ for $j = 1, \ldots, d^n - 1$. Thus

$$\delta^{d^n} \leq \prod |z_j| = |1 - \lambda^n|.$$

We now construct a $\lambda$ for which this is impossible, thus contradicting the existence of $h$. Suppose $q_1 < q_2 < \cdots$ is an increasing sequence of integers, and set $\theta = \sum_{k=1}^\infty 2^{-q_k}$ and $\lambda = \exp(2\pi i \theta)$. Then

$$|1 - \lambda^{2^{q_k}}| \sim 2^{q_k - q_{k+1}}.$$

Taking logarithms, we find that

$$q_{k+1} \leq C(\delta) d^{2^{q_k}}.$$

If we define inductively $q_k$'s to grow very rapidly, say with $\log q_{k+1} \geq k2^{q_k}$, this inequality is violated for any $d$ and $\delta > 0$. □

An example as above was first given in 1917 by G.A. Pfeiffer. The work was continued by H. Cremer, who proved in 1938 that if $|\lambda| = 1$ and $\liminf |\lambda^n - 1|^{1/n} = 0$, then there is an analytic function $f(z) = \lambda z + \cdots$ such that the Schröder equation (6.1) has no solution. In a seminal paper in 1942, C.L. Siegel gave the first example of a unimodular $\lambda$ for which (6.1) is solvable.

A real number $\theta$ is *Diophantine* if it is badly approximable by rational numbers, in the sense that there exist $c > 0$ and $\mu < \infty$ so that

$$\left| \theta - \frac{p}{q} \right| \geq \frac{c}{q^\mu} \tag{6.2}$$

for all integers $p$ and $q$, $q \neq 0$. This occurs if and only if $\lambda = e^{2\pi i \theta}$ satisfies

$$|\lambda^n - 1| \geq cn^{1-\mu}, \qquad n \geq 1.$$

For fixed $\mu > 2$, the condition (6.2) holds for a.e. real number $\theta$. Indeed, if $E$ is the set of $\theta \in [0, 1]$ such that $|\theta - p/q| < q^{-\mu}$ infinitely often, then the measure of $E$ is estimated in the obvious way by

$$|E| \leq \sum_{q=n}^{\infty} 2 \cdot q^{-\mu} \cdot q = \mathcal{O}(n^{2-\mu}) \to 0.$$

In particular, almost all real numbers are Diophantine. By a theorem of J. Liouville [HaW], any algebraic surd $\theta$ of degree $m$ satisfies (6.2) for $\mu > m$.

THEOREM 6.4 (Siegel). *If $\theta$ is Diophantine, and if $f$ has fixed point at 0 with multiplier $e^{2\pi i \theta}$, then there exists a solution to the Schröder equation (6.1), that is, $f$ can be conjugated near 0 to multiplication by $e^{2\pi i \theta}$.*

*Proof.* We want to solve $h(\lambda z) = f(h(z))$. If we define $\hat{f}$ and $\hat{h}$ by $f(z) = \lambda z + \hat{f}(z)$ and $h(z) = z + \hat{h}(z)$, then the equation can be written

$$\hat{h}(\lambda z) - \lambda \hat{h}(z) = \hat{f}(h(z)). \tag{6.3}$$

Siegel's original method was to expand both sides in power series using

$$\hat{h}(z) = \sum_{n=2}^{\infty} a_n z^n, \qquad \hat{f}(z) = \sum_{n=2}^{\infty} b_n z^n$$

to inductively obtain equations of the form $a_n(\lambda^n - \lambda) = A_n(a_2, \cdots,$ $a_{n-1}, b_2, \cdots, b_n)$, $n \geq 2$. Using $|\lambda^n - \lambda| > cn^{1-\mu}$ he was able to estimate the $a_n$'s and prove the power series converges. This is not easy. We take a different route, following the proof in [SiM].

We shall use KAM theory (A.N. Kolmogorov, V.I. Arnold, J. Moser) in a simple case. We consider coordinate changes $\psi$ around $z = 0$, normalized so that $\psi(z) = z + \hat{\psi}(z)$, where $\hat{\psi}(z) = \mathcal{O}(z^2)$. Instead of finding a solution to (6.1) right away, we build a $\psi$ so that

$$\psi^{-1} \circ f \circ \psi = g(z) = \lambda z + \hat{g}(z), \qquad (6.4)$$

where $\hat{g}$ is smaller than $\hat{f}$ in some sense. (Note that $\hat{g} = 0$ solves the problem.) We then repeat the procedure with $f$ replaced by $g$ and $\psi$ defined on a slightly smaller disk. Careful estimates show we obtain the desired $h$ in the limit on a disk of positive radius.

So suppose $\hat{f}(z)$ is as above. The idea is to replace (6.3) by its linear version, by replacing $h$ in the right-hand side by $z$,

$$\hat{\psi}(\lambda z) - \lambda\hat{\psi}(z) = \hat{f}(z). \qquad (6.5)$$

This can easily be solved for $\hat{\psi}$,

$$\hat{\psi}(z) = \sum_{j=2}^{\infty} \frac{b_j}{\lambda^j - \lambda} z^j,$$

and our hope is that the corresponding $\hat{g}$ defined by (6.4) is smaller than $\hat{f}$. We estimate $\hat{g}$ and $\hat{g}'$ using the assumptions

$$\frac{1}{|\lambda^n - 1|} \leq c_0 \frac{n^\mu}{\mu!},$$

$$|\hat{f}'(z)| < \delta \quad \text{in} \quad \Delta(0, r).$$

We have two parameters $\delta, r$ and want to find the corresponding values for $g = \psi^{-1} \circ f \circ \psi$. First we estimate $\hat{\psi}$ in a slightly smaller disk $\Delta(0, r(1-\eta))$ for some $0 < \eta < 1/5$. From the Cauchy estimates for the power series coefficients of $\hat{f}'$, we have

$$|b_j| \leq \frac{\delta}{jr^{j-1}}.$$

and so for $z \in \Delta(0, (1 - \eta)r)$, we have

$$
\begin{aligned}
|\hat{\psi}'| &\leq \sum_{j=2}^{\infty} \frac{j b_j}{|\lambda^j - \lambda|} r^{j-1} (1 - \eta)^{j-1} \leq \frac{c_0 \delta}{\mu!} \sum_{j=2}^{\infty} j^{\mu} (1 - \eta)^{j-1} \\
&< \frac{c_0 \delta}{\mu!} \sum_{j=1}^{\infty} j(j+1) \cdots (j + \mu - 1)(1 - \eta)^{j-1} = \frac{c_0 \delta}{\eta^{\mu+1}}.
\end{aligned}
$$

If we also assume $c_0 \delta < \eta^{\mu+2}$ then we have $|\hat{\psi}'| \leq \eta$ in this disk.

Clearly $|\hat{\psi}'| \leq \eta$ implies that $\psi$ maps $\Delta(0, r(1 - 4\eta))$ into $\Delta(0, r(1 - 3\eta))$. Furthermore $\psi$ takes every value in $\Delta(0, r(1 - 2\eta))$ precisely once in $\Delta(0, r(1 - \eta))$, in view of the argument principle and the observation that $|\psi(z)| \geq r(1 - 2\eta)$ for $z \in \partial\Delta(0, r(1 - \eta))$ and $\psi(z) = 0$ only at 0 in this disk. Now consider how $g = \psi^{-1} \circ f \circ \psi$ maps $\Delta(0, r(1 - 4\eta))$. First $\psi$ takes this disk into $\Delta(0, r(1 - 3\eta))$, then $f$ takes this to $\Delta(0, r(1 - 2\eta))$ if $\delta < \eta$, and finally $\psi^{-1}$ takes $\Delta(0, r(1 - 2\eta))$ into $\Delta(0, r(1 - \eta))$. Thus $g$ maps $\Delta(0, r(1 - 4\eta))$ into $\Delta(0, r(1 - \eta))$.

We now estimate $\hat{g}$. From (6.4) we have

$$
\hat{g}(z) + \hat{\psi}(\lambda z + \hat{g}) = \lambda \hat{\psi}(z) + \hat{f}(z + \hat{\psi}),
$$

and using (6.5) we obtain

$$
\hat{g}(z) = \hat{\psi}(\lambda z) - \hat{\psi}(\lambda z + \hat{g}) + \hat{f}(z + \hat{\psi}) - \hat{f}(z).
$$

We wish to estimate $\hat{g}$ in $\Delta(0, r(1 - 4\eta))$. Let $C$ be the maximum of $|\hat{g}|$ in this disk. Then

$$
C \leq \sup |\hat{\psi}'| \cdot C + \sup |\hat{f}(z + \hat{\psi}) - \hat{f}(z)| \leq \eta C + \delta \sup |\hat{\psi}|.
$$

Since $\hat{\psi}(z) = \mathcal{O}(z^2)$, we obtain from the Schwarz lemma and an integration

$$
|\hat{\psi}(z)| \leq \frac{1}{2} \frac{c_0 \delta}{\eta^{\mu+1}} (1 - \eta)r, \qquad |z| \leq (1 - \eta)r.
$$

Substituting and solving for $C$, we obtain $C \leq c_0 \delta^2 \eta^{-(\mu+1)} r / 2$. Cauchy's estimate, applied to a disk of radius $r\eta$, then yields

$$
|\hat{g}'| \leq \frac{1}{2} \frac{c_0 \delta^2}{\eta^{\mu+2}}, \qquad |z| \leq r(1 - 5\eta).
$$

Note in particular that the $\delta$ in the estimate for $\hat{f}'$ has become a $\delta^2$ in the estimate for $\hat{g}'$. This improvement is what allows the iteration to work.

Consider what we have done. We have taken an $f$ satisfying $|\hat{f}'| \leq \delta$ in $\Delta(0, r)$ and replaced it by a $g$ satisfying $|\hat{g}'| \leq c_0 \delta^2 \eta^{-(\mu+2)}/2$ in $\Delta(0, r(1 - 5\eta))$. To do this, we needed to assume $r \leq r_0$, where $f$ is defined on $\Delta(0, r_0)$, and also

$$0 < \eta < 1/5, \quad c_0 \delta < \eta^{\mu+2}, \quad \delta < \eta.$$

If we take $c_1 > 0$ small enough and require $\eta \leq c_1$, then the first condition is satisfied, and the third condition follows from the second.

Suppose now $\eta_0, \delta_0$ have been chosen to satisfy these conditions. Define $\alpha$ so that $\alpha^{\mu+2} = 1/2$, and define sequences by $\eta_n = \alpha^n \eta_0$, $r_{n+1} = r_n(1 - 5\eta_n)$, and $\delta_{n+1} = c_0 \delta^2 \eta_n^{-(\mu+2)}/2$. These satisfy the required condition $c_0 \delta_n \leq \eta_n^{\mu+2}$, as can be checked by induction. We have now also constructed sequences $\{\psi_n\}$ and $\{g_n\}$ with $g_0 = f$ and $g_n = \psi_n^{-1} \circ g_{n-1} \circ \psi_n$, or equivalently

$$g_n = \psi_n^{-1} \circ \psi_{n-1}^{-1} \circ \cdots \circ \psi_1^{-1} \circ f \circ \psi_1 \circ \psi_2 \circ \cdots \circ \psi_n.$$

Set $R = r_0 \prod(1 - 5\eta_n) > 0$, then $|\hat{g}_n'| \leq \delta_n \sim \eta_n^{\mu+2} \to 0$ on $\Delta(0, R)$, so $g_n \to \lambda z$ on the disk. Thus $\{\psi_1 \circ \cdots \circ \psi_n\}$ converges to a mapping $h$ which conjugates $f$ to $\lambda z$, as desired. $\square$

For quadratic polynomials $P(z) = e^{2\pi i \theta} z + z^2$, precise conditions are known for the existence of a conjugation. Such exists if and only if

$$\sum_{n=1}^{\infty} \frac{\log q_{n+1}}{q_n} < \infty,$$

where $\{p_n/q_n\}$ is the sequence of rational approximations to $\theta$ coming from its continued fraction expansion. The sufficiency of this condition was proved by A.D. Brjuno (1965), and the necessity was established by J.-C. Yoccoz (1988). In Section V.1 we give an elementary proof, due to Yoccoz, of the existence of the conjugation for almost all $\theta$.

## 7.  Homeomorphisms of the Circle

Any orientation preserving homeomorphism $f$ of the unit circle can be expressed in the form $f(t) = e^{2\pi i F(t)}$, where $F$ is an increasing homeomorphism of the real line $\mathbf{R}$ that satisfies $F(t+1) = F(t)+1$. The lift $F$ of $f$ is unique up to adding an integer. We wish to assign a rotation number to $f$, which will measure the average advance of $F$ over an interval of length 1, that is, the average speed of $F$.

LEMMA 7.1. *Let $F$ be as above, and suppose $F(0) > 0$. Fix $p \geq 1$. If $m = m(p)$ is the first integer such that $F^m(0) > p$, then*

$$\frac{p}{m} \leq \liminf_{n \to \infty} \frac{F^n(0)}{n} \leq \limsup_{n \to \infty} \frac{F^n(0)}{n} \leq \frac{p}{m} + \frac{1 + F(0)}{m}.$$

*Proof.* If $0 \leq t \leq 1$, then $F(0) \leq F(t) \leq F(0)+1$. If $s > 0$ is arbitrary, and $j$ is the integral part of $s$, then $s + F(0) - 1 \leq j + F(0) = F(j) \leq F(s) \leq F(j+1) = F(0)+j+1 \leq s+F(0)+1$. We check by induction that

$$s + k(F(0) - 1) \leq \; F^k(s) \leq s + k(F(0) + 1), \quad s > 0, \; k \geq 1,$$
$$F^{k(m-1)}(0) \leq \; kp \leq F^{km}(0), \qquad\qquad k \geq 1.$$

Write $n = km + q$ where $0 \leq q < m$. Then $kp + q(F(0) - 1) \leq F^q(kp) \leq F^q(F^{km}(0)) = F^n(0) = F^{q+k}(F^{k(m-1)}(0)) \leq F^{q+k}(kp) \leq kp + (q + k)(1 + F(0))$. This gives

$$\frac{kp}{n} + \frac{q(F(0) - 1)}{n} \leq \frac{F^n(0)}{n} \leq \frac{kp}{n} + \frac{k + q}{n}(1 + F(0)).$$

Since $k/n \to 1/m$ and $q/n \to 0$ as $n \to \infty$, the lemma follows. $\square$

There is a similar estimate if $F(0) < 0$. It follows from the estimate that

$$\alpha(F) = \lim_{n \to \infty} \frac{F^n(0)}{n}$$

exists. We call $\alpha(F)$ the *rotation number* of $F$. We define the *rotation number* $\alpha(f)$ of $f$ to be the residue class of $\alpha(F)$ modulo 1. This is independent of the lift $F$.

One checks that the translation $t \to t + \theta$ has rotation number $\theta$, so the rotation $f_\theta(\zeta) = e^{2\pi i \theta}\zeta$ has rotation number $\theta$ (mod 1). From the estimates in the lemma it is clear that the rotation number of

$F$ depends continuously on any parameters on which $F$ depends continuously, as does the rotation number of $f$.

We have $F^n(t)/n \to \alpha(F)$ for any real number $t$. This follows from the periodicity modulo 1 of $F$ and the estimate $F^n(0) \le F^n(t) \le F^n(0) + 1$ for $0 \le t \le 1$.

If $F(t) \ge t + \alpha_0 + \varepsilon$ for $0 \le t \le 1$, then $F^n(0) \ge n(\alpha_0 + \varepsilon)$, and $\alpha(F) \ge \alpha_0 + \varepsilon$. Similarly, if $F(t) \le t + \alpha_0 - \varepsilon$ for $0 \le t \le 1$, then $\alpha(F) \le \alpha_0 - \varepsilon$. It follows that $F(t) - t$ assumes the value $\alpha(F)$ somewhere on the interval $0 \le t \le 1$. In particular, if $\alpha(F) = 0$, then there is a fixed point for $F$. Conversely, if $F$ has a fixed point $t_0$, then $\alpha(F) = \lim F^n(t_0)/n = 0$. Similarly, $f$ has a fixed point if and only if $\alpha(f) \equiv 0$, and since $\alpha(f^n) = n\alpha(f)$, $f$ has a periodic point if and only if $\alpha(f)$ is rational.

In general, $\alpha(F \circ G) \ne \alpha(F) + \alpha(G)$. Nevertheless, $\alpha(F)$ is invariant under conjugation. Indeed, suppose $\Phi$ satisfies the same conditions as $F$. If $k(n)$ is the integral part of $F^n(0)$, then $\Phi(F^n(0)) = \Phi(F^n(0) - k(n)) + k(n)$, and

$$
\begin{aligned}
\alpha(\Phi \circ F \circ \Phi^{-1}) &= \lim \frac{(\Phi \circ F \circ \Phi^{-1})^n(\Phi(0))}{n} = \lim \frac{\Phi(F^n(0))}{n} \\
&= \lim \frac{k(n)}{n} = \alpha(F).
\end{aligned}
$$

Thus $\alpha(f)$ is also invariant under conjugation.

Rotation numbers were first introduced by H. Poincaré (1885) in connection with ordinary differential equations. A. Denjoy (1932) showed that if $f$ has irrational rotation number and $f'$ has bounded variation, then $f$ is conjugate to the corresponding irrational rotation. V.I. Arnold (1961) showed that if $f$ is analytic with Diophantine rotation number $\alpha(f)$, and if $f$ is a small perturbation of a rotation, then the conjugating map can be taken to be analytic in a neighborhood of the circle. Arnold conjectured the global version of his theorem (no restriction on being close to a rotation). Arnold's conjecture was proved for a large class of Diophantine numbers by M. Herman (1979), and extended to cover all Diophantine numbers by Yoccoz (1984). We give a proof of Arnold's theorem (which came to us through a lecture of T. Kuusalo), since it is very similar to that of Theorem 6.4, and since we will use the theorem later.

THEOREM 7.2 (Arnold). *Let $\alpha$ be Diophantine, and let $\sigma > 1$. Then there is $\varepsilon > 0$ such that if $f$ is any homeomorphism of the circle* with rotation number $\alpha(f) \equiv \alpha$, which extends to be analytic and univalent on the annulus $\{1/\sigma < |w| < \sigma\}$ and satisfies $|f(w) - e^{2\pi i \alpha}w| < \varepsilon$ there, then $f$ is conformally conjugate to the rotation $e^{2\pi i \alpha}w$ on the annulus $\{1/\sqrt{\sigma} < |w| < \sqrt{\sigma}\}$.

*Proof.* It will be convenient to work with the lifts of functions to a strip rather than functions on an annulus. Let $S_\rho$ be the horizontal strip $\{x + iy : |y| < \rho\}$. For $g(z)$ defined on $S_\rho$ let $\|g\|_\rho$ denote the supremum of $|g(z)|$ over $S_\rho$. Replacing $f$ by its lift, we assume $f(z)$ is analytic on $S_\rho$ and satisfies $f(z + 1) = f(z) + 1$. We may assume that $f(x)$ is increasing on $\mathbf{R}$. Write $f(z) = z + \alpha + \hat{f}(z)$. Then $\hat{f}$ is periodic and has a Laurent expansion

$$\hat{f}(z) = \sum_{n=-\infty}^{+\infty} b_n e^{2\pi i n z}, \qquad e^{-2\pi\rho} < |e^{2\pi i z}| < e^{2\pi\rho}.$$

We wish to find a univalent function $h$ on $S_{\rho/2}$ that satisfies $h^{-1} \circ f \circ h(z) = z + \alpha$, that is, $h(z + \alpha) - h(z) = \alpha + \hat{f}(h(z))$. As before, we consider the linearized version

$$\psi(z + \alpha) - \psi(z) = \alpha + \hat{f}(z). \tag{7.1}$$

In Theorem 6.4 the multiplier was just the coefficient of $z$ in the series expansion, the term corresponding to $\alpha + b_0$ here. To simplify matters we assume for the moment that $b_0 = 0$, that is, the constant term of $f$ is just the rotation number. The linearized equation is then solved by $\psi(z) = z + \hat{\psi}(z)$, where

$$\hat{\psi}(z) = \sum_{n \neq 0} \frac{b_n}{\lambda^n - 1} e^{2\pi i n z}, \qquad \lambda = e^{2\pi i \alpha}.$$

We assume that

$$\frac{1}{|\lambda^n - 1|} \leq c_0 |n|^{\mu - 1}, \qquad n \neq 0.$$

From the Cauchy estimates for the Laurent series coefficients of $\hat{f}$ we have

$$|b_n| \leq \|\hat{f}\|_\rho e^{-2\pi |n| \rho}.$$

Summing the series gives a constant $C_1$ depending only on $c_0$ and $\mu$ such that

$$\|\hat{\psi}\|_{\rho - \sigma} \leq C_1 \|\hat{f}\|_\rho \sigma^{-\mu}, \qquad 0 < \sigma \leq 1. \tag{7.2}$$

Now suppose $b_0$ is arbitrary. Let $\hat{f}_0(z) = \hat{f}(z) - b_0$ and suppose $\psi_0(z) = z + \hat{\psi}_0(z)$ is the function for $z + \alpha + \hat{f}_0(z)$ obtained by the above procedure. Then $\psi_0$ and $\hat{f}_0$ satisfy (7.1), so that

$$\hat{\psi}_0(z + \alpha) - \hat{\psi}_0(z) = \hat{f}(z) - b_0. \qquad (7.3)$$

Since $|b_0| \leq \|\hat{f}\|_\rho$, we have $\|\hat{f}_0\|_\rho \leq 2\|\hat{f}\|_\rho$ and so (7.2) holds for $\hat{\psi}_0$ providing $C_1$ is doubled.

We start now with $f$ satisfying $\|\hat{f}\|_\rho \leq \delta$. Let $0 < \eta < 5\rho$. As before we see that if $\delta < \eta$ and $\delta < \eta^{\mu+1}/C_1$, then $\psi_0(S_{\rho-4\eta}) \subset S_{\rho-3\eta}$, $f(S_{\rho-3\eta}) \subset S_{\rho-2\eta}$, and $\psi_0(S_{\rho-\eta}) \supset S_{\rho-2\eta}$, so that

$$g(z) = \psi_0^{-1} \circ f \circ \psi_0(z) = z + \alpha + \hat{g}(z)$$

is defined on $S_{\rho-4\eta}$ and has range in $S_{\rho-\eta}$. Also, $\|\hat{g}\|_{\rho-4\eta} \leq 3\eta$. We wish to obtain an estimate of the form $\|\hat{g}\|_{\rho-4\eta} \leq C_0\delta^2\eta^{-\mu-1}$. The proof is then completed as before by iterating, with $\delta^2$ providing the margin of victory.

Equating the expressions

$$\begin{aligned} \psi_0(g(z)) &= g(z) + \hat{\psi}_0(g(z)) = z + \alpha + \hat{g}(z) + \hat{\psi}_0(z + \alpha + \hat{g}(z)), \\ f(\psi_0(z)) &= \psi_0(z) + \alpha + \hat{f}(\psi_0(z)) = z + \hat{\psi}_0(z) + \alpha + \hat{f}(z + \hat{\psi}_0(z)), \end{aligned}$$

we find using (7.3) that

$$\hat{g}(z) = [\hat{f}(z + \hat{\psi}_0(z)) - \hat{f}(z)] - [\hat{\psi}_0(z + \alpha + \hat{g}(z)) - \hat{\psi}_0(z + \alpha)] + b_0.$$

Since $\alpha$ is also the rotation number of $g$, $g(x)$ is equal to $x + \alpha$ for some $x$, so that $\hat{g}(x) = 0$ for some $x \in \mathbf{R}$. Thus $b_0$ is estimated by the terms in square brackets, and it suffices to estimate these. From the Schwarz inequality and (7.2) we obtain

$$|\hat{f}(z + \hat{\psi}_0(z)) - \hat{f}(z)| \leq \frac{2\delta}{\eta}\|\hat{\psi}_0\|_{\rho-4\eta} \leq C_2\delta^2\eta^{-\mu-1}.$$

Set $C = \|\hat{g}\|_{\rho-4\eta}$. Then $C \leq 3\eta$ and

$$|\hat{\psi}_0(z + \alpha + \hat{g}(z)) - \hat{\psi}_0(z + \alpha)| \leq 2C_1\delta\eta^{-\mu}\frac{C}{3\eta} < C_1\delta\eta^{-\mu-1}C.$$

Hence $C \leq 2[C_2\delta^2\eta^{-\mu-1} + C_1\delta\eta^{-\mu-1}C]$. If $\delta < \eta^{\mu+1}/(4C_1)$, we can solve for $C$ as before and get the desired estimate.

For the iteration procedure the essential condition on $\delta$ has the form $\delta < c_1 \eta^{\mu+1}$, where $c_1 < 1$. We choose $\eta_1$ to be the smaller of $\rho/16$ and $1/16$, and we set $\varepsilon = \delta_1 = c_1 \eta_1^{\mu+1}$, which depends only on $\rho$ and $\alpha$. Setting $\eta_n = \eta_1/2^{n-1}$, $\rho_n = \rho - 4(\eta_1 + \cdots + \eta_{n-1})$, and $\delta_n = \delta_{n-1}^{3/2}$, we obtain for $f$ with $\|\hat{f}\|_\rho < \varepsilon$ a sequence of conjugations $\{\psi_n\}$ on $S_{\rho_n}$ and a sequence $g_n = \psi_n^{-1} \circ \ldots \circ \psi_1^{-1} \circ f \circ \psi_1 \circ \ldots \circ \psi_n$. By induction the corresponding $\hat{g}_n$'s satisfy $\|\hat{g}_n\|_{\rho_n} \leq \delta_n$. Then $g_n$ converges uniformly to $z + \alpha$ on $S_{\rho/2}$, and $\psi_1 \circ \cdots \circ \psi_n$ converges on $S_{\rho/2}$ to a univalent mapping $h$ which conjugates $f$ to translation by $\alpha$. $\square$

# III

# Basic Rational Iteration

We consider a rational mapping $R(z)$ and use it to partition the sphere into two disjoint invariant sets, on one of which $R(z)$ is well-behaved (the Fatou set), on the other of which $R(z)$ has chaotic behavior (the Julia set). The first milestone of the theory is a theorem of Fatou and Julia that the repelling periodic points are dense in the Julia set. From this follows the homogeneous nature of the Julia set. In the words of Julia, "*la structure de $E'$ est la même dans toutes ses parties.*"

## 1. The Julia Set

Let $R(z)$ be rational, and assume always that $R$ has degree $d \geq 2$. Thus $R = P/Q$ where $P, Q$ are polynomials with no common factors and $d = \max(\deg P, \deg Q) \geq 2$. Then $R(z)$ is a $d$-fold branched covering of the Riemann sphere, and in fact every $d$-to-1 conformal, branched covering of $\overline{\mathbf{C}}$ comes from some such $R$. We define the iterates of $R$ as before, $R^2 = R \circ R$, $R^n = R^{n-1} \circ R$ and note that $R^n$ has degree $d^n$.

Given $z_0$, the sequence of points $\{z_n\}$ defined by $z_n = R(z_{n-1})$ is called the *(forward) orbit* of $z_0$. The chain rule gives the useful

identity
$$(R^n)'(z_0) = R'(z_0)R'(z_1) \cdots R'(z_{n-1}).$$

The point $z_0$ is called *periodic* if $z_n = z_0$ for some $n$. This occurs if and only if $z_0$ is a fixed point for $R^n$. The minimal $n$ is its *period*. The orbit $\{z_1, z_2, ..., z_n = z_0\}$ is called a *cycle*. We classify the cycle as *attracting*, *repelling*, *rationally neutral* or *irrationally neutral* according to the type of fixed point for $R^n$. Thus for instance the cycle is attracting if and only if $\prod |R'(z_j)| < 1$, and this does not depend on the choice of point in the cycle. The point $z_0$ is called *preperiodic* if $z_k$ is periodic for some $k > 0$, and *strictly preperiodic* if it is preperiodic but not periodic.

Recall that a *critical point* of $R$ is a point on the sphere where $R$ is not locally one-to-one. These consist of solutions of $R'(z) = 0$ and of poles of $R$ of order two or higher. The images of the critical points are the *critical values* of $R$. The *order* of a critical point $z$ is the integer $m$ such that $R$ is $(m+1)$-to-1 in a punctured neighborhood of $z$. If $z$ is not a pole this is its multiplicity as a zero of $R'$. There are $2d - 2$ critical points, counting multiplicity. One way to see this is to compose $R$ with a Möbius transformation, to reduce to the case where $\infty$ is neither a critical point nor a critical value, and $R(\infty) = 0$. Then
$$R(z) = \frac{P(z)}{Q(z)} = \frac{\alpha z^{d-1} + \cdots}{\beta z^d + \cdots},$$
where $\alpha, \beta \neq 0$, and so
$$R'(z) = \frac{Q(z)P'(z) - P(z)Q'(z)}{Q(z)^2} = \frac{-\alpha\beta z^{2d-2} + \cdots}{Q(z)^2}.$$

The *Fatou set* $\mathcal{F}$ of $R$ is defined to be the set of points $z_0 \in \overline{\mathbf{C}}$ such that $\{R^n\}$ is a normal family in some neighborhood of $z_0$. The *Julia set* $\mathcal{J}$ is the complement of the Fatou set. Thus $z_0 \in \mathcal{F}$ if and only if the family $\{R^n\}$ is equicontinuous on some neighborhood of $z_0$, with respect to the spherical metric of $\overline{\mathbf{C}}$. Evidently the Fatou set is an open subset of $\overline{\mathbf{C}}$, and the Julia set is compact.

If $R$ has an attracting fixed point $z_0$, then the basin of attraction $A(z_0)$ is contained in the Fatou set. On the other hand, since the iterates of $R$ converge to $z_0$ on $A(z_0)$ but not on its complement, the iterates cannot be normal on any open set meeting $\partial A(z_0)$, and $\partial A(z_0)$ is included in the Julia set. We will see later (Theorem 2.1) that $\mathcal{J} = \partial A(z_0)$.

EXAMPLE. Consider $R(z) = z^2$. Then $R^n(z) = z^{2^n}$ converges to $0$ in $\{|z| < 1\}$ and to $\infty$ in $\{|z| > 1\}$. The Julia set is the unit circle $\{|z| = 1\}$.

EXAMPLE. Let $R(z) = z^2 - 2$. We have seen in Section II.1 that the basin of attraction of $\infty$ coincides with $\overline{\mathbf{C}}\backslash[-2, 2]$, so the Julia set $\mathcal{J}$ is the closed interval $[-2, 2]$.

The preceding two examples may be misleading. Julia sets are almost always intricate fractal sets, which are conformally self-similar. See the figures in Chapter VIII of the Julia sets of $R(z) = z^2 + c$ for various other values of $c$.

EXAMPLE. Take $\lambda$, $|\lambda| = 1$, such that $R(z) = \lambda z + z^2$ is conjugate to an irrational rotation $\lambda z$ near $0$ (such exist by Theorem II.6.4). Then $\{R^n\}$ is normal in neighborhoods of $0$ and $\infty$, so there are components $U_0$ and $U_\infty$ of $\mathcal{F}$ containing these points. Since $\infty$ is an attracting fixed point and $R^n(0) \not\to \infty$, they must be distinct. Hence $R^n(U_0)$ never hits $U_\infty$, and $\{R^n\}$ is uniformly bounded on $U_0$. Also $U_0$ is simply connected, since by the maximum principle $\{R^n\}$ is bounded within any closed curve in $U_0$. In view of Theorem II.6.2, we see that $U_0$ is characterized as the largest domain containing $0$, invariant under $R$, on which the conjugation map exists.

A simply connected component of the Fatou set in which $R$ is conjugate to an irrational rotation is called a *Siegel disk*. An example is illustrated in Section V.1.

THEOREM 1.1. *The Julia set $\mathcal{J}$ contains all repelling fixed points and all neutral fixed points that do not correspond to Siegel disks. The Fatou set $\mathcal{F}$ contains all attracting fixed points and all neutral fixed points corresponding to Siegel disks.*

*Proof.* This is obvious, in view of Theorem II.6.2. □

In the third example above we saw that a disk around $0$ was in the Fatou set for some $|\lambda| = 1$, but if $|\lambda| > 1$ then it is a repelling fixed point so is in the Julia set. Thus $\mathcal{F}$ and $\mathcal{J}$ may move quite discontinuously as $R$ varies. We shall see other examples of this later.

THEOREM 1.2. *The Julia set $J$ is nonempty.*

*Proof.* Recall our standing assumption that $R$ has degree $d \geq 2$. Suppose $J = \emptyset$. Then $\{R^n\}$ is a normal family on all of $\overline{C}$, and so there is a subsequence $\{n_j\}$ such that $R^{n_j}(z) \to f(z)$ for some analytic function $f$ from $\overline{C}$ to $\overline{C}$. Since $f$ is analytic on all of $\overline{C}$ it is a rational function. If $f$ is constant then the image of $R^{n_j}$ is eventually contained in a small neighborhood of the constant value, which is impossible since $R^n$ covers $\overline{C}$. If $f$ is not constant, eventually $R^{n_j}$ has the same number of zeros as $f$ (apply the argument principle), which is also impossible since $R^n$ has degree $d^n$. $\square$

We say that a set $E$ is *completely invariant* if both it and its complement are invariant. Since $R$ is onto, this occurs if and only if $R^{-1}(E) = E$.

THEOREM 1.3. *The Julia set $J$ is completely invariant.*

*Proof.* Clearly $R^{-1}(F) \subseteq F$. Suppose $z_0 \in F$, and suppose $\{R^{n_j+1} = R^{n_j} \circ R\}$ converges uniformly (in the spherical metric) on a neighborhood of $z_0$. Since $R$ maps open neighborhoods of $z_0$ onto open neighborhoods of $R(z_0)$, $\{R^{n_j}\}$ converges uniformly on a neighborhood of $R(z_0)$. It follows that $R(z_0) \in F$, and $R(F) \subseteq F$. Hence $F$ is completely invariant, as is its complement $J$. $\square$

THEOREM 1.4. *For any $N \geq 1$, the Julia set of $R$ coincides with that of $R^N$.*

*Proof.* The Fatou sets for $R$ and $R^N$ are the same, since $\{R^n\}$ is normal on an open set $U$ if and only if $\{R^{nN}\}$ is normal on $U$. $\square$

Let $z \in J$ and let $U$ be any neighborhood of $z$. By Montel's theorem (Theorem I.3.2), the sequence $\{R^n\}$ on any such $U$ omits a set $E_z$ containing at most two points.

THEOREM 1.5. *The set $E_z$ is independent of $z$ (so shall be denoted as $E$). If $E$ is a singleton, we can conjugate it to $\infty$, and then $R(z)$ is a polynomial. If $E$ consists of two points, we can conjugate these to $0$ and $\infty$, and then $R(z)$ is $Cz^d$ or $Cz^{-d}$. In all cases, $E$ is contained in the Fatou set $F$.*

*Proof.* By the definition $R^{-1}(E_z) \subset E_z$. If $E_z$ is one point $a$, then $R(a) = a$. Take $a = \infty$. Since $R^{-1}(\infty) = \infty$, there are no other poles, and $R$ is a polynomial. Clearly $E_z$ is independent of $z$. If $E_z$ consists of two points, we may assume these to be $0, \infty$ and either $R(0) = 0$, $R(\infty) = \infty$ or $R(0) = \infty$, $R(\infty) = 0$. In the first case, $R$ is a polynomial with 0 as its only zero, so $R(z) = Cz^d$. Similarly, $R(z) = Cz^{-d}$ in the second case. $\square$

The set $E$ is called the *exceptional set* of $R$. It follows immediately from the definition of $E$ that if $z \notin E$ then $\mathcal{J}$ is adherent to the inverse orbit $\cup_{n \geq 1} R^{-n}(z)$ of $z$. Since $E$ is disjoint from $\mathcal{J}$, we obtain in particular the next two theorems as corollaries.

THEOREM 1.6. *The backward iterates of any $z \in \mathcal{J}$ are dense in $\mathcal{J}$.*

This theorem can be used as a basis for computing $\mathcal{J}$ (the *inverse iteration method*). It is effective when the degree of $R$ is small, so that the number of inverse iterates does not grow uncontrollably fast.

THEOREM 1.7. *Any nonempty completely invariant subset of $\mathcal{J}$ is dense in $\mathcal{J}$. If $D$ is a union of components of $\mathcal{F}$ that is completely invariant, then $\mathcal{J} = \partial D$.*

*Proof.* For the second statement, note that $\partial D$ is a completely invariant subset of $\mathcal{J}$. $\square$

THEOREM 1.8. *The Julia set $\mathcal{J}$ contains no isolated points, that is, $\mathcal{J}$ is a perfect set.*

*Proof.* Take $z_0 \in \mathcal{J}$ and $U$ an open neighborhood of $z_0$. First assume $z_0$ is not periodic and choose $z_1$ with $R(z_1) = z_0$. Then $R^n(z_0) \neq z_1$ for all $n$. Since $z_1 \in \mathcal{J}$, backward iterates of $z_1$ are dense in $\mathcal{J}$, so there is a $\zeta \in U$ with $R^m(\zeta) = z_1$. Thus $\zeta \in \mathcal{J} \cap U$ and $\zeta \neq z_0$.

Next suppose $R^n(z_0) = z_0$ for some minimal $n$. If $z_0$ were the only solution of $R^n(z) = z_0$ then $z_0$ would be a superattracting fixed point for $R^n$, contradicting $z_0 \in \mathcal{J}$. Hence there is $z_1 \neq z_0$ with $R^n(z_1) = z_0$. Furthermore $R^j(z_0) \neq z_1$ for all $j$ since otherwise it would hold for some $0 \leq j < n$ (by periodicity) and hence $R^j(z_0) = R^{n+j}(z_0) = R^n(z_1) = z_0$, contradicting the minimality of $n$. As before, $z_1$ must have a preimage in $U \cap \mathcal{J}$ which cannot be $z_0$. $\square$

EXAMPLE. The Julia set of a Blaschke product $B(z)$ of degree $d \geq 2$ is either the unit circle $\partial \Delta$ or a Cantor set on $\partial \Delta$. Indeed, the iterates of $B(z)$ are normal on $\Delta$ and outside $\overline{\Delta}$, so $\mathcal{J}$ is a perfect subset of $\partial \Delta$. Suppose some $z_0 \in \mathcal{J}$ is adherent to $\mathcal{F} \cap \partial \Delta$. Since $\mathcal{F} \cap \partial \Delta$ is completely invariant, and since $\arg B^n$ is strictly increasing on $\partial \Delta$, each inverse iterate of $z_0$ is also adherent to $\mathcal{F} \cap \partial \Delta$. By Theorem 1.6, $\mathcal{J}$ is adherent to $\mathcal{F} \cap \partial \Delta$, consequently $\mathcal{J}$ is totally disconnected.

THEOREM 1.9. *If the Julia set $\mathcal{J}$ has nonempty interior, then $\mathcal{J}$ coincides with the extended complex plane $\overline{\mathbf{C}}$.*

*Proof.* Suppose there is an open $U \subset \mathcal{J}$. Then $R^n(U) \subset \mathcal{J}$. But $\cup R^n(U) = \overline{\mathbf{C}} \backslash E$ is dense in $\overline{\mathbf{C}}$ and since $\mathcal{J}$ is closed, $\mathcal{J} = \overline{\mathbf{C}}$. $\square$

EXAMPLE. The Lattès function $(z^2 + 1)^2/4z(z^2 - 1)$ discussed in Section II.1 has Julia set $\mathcal{J} = \overline{\mathbf{C}}$, since a dense subset of $\overline{\mathbf{C}}$ is iterated to the repelling fixed point at $\infty$. We will see later, as a consequence of Theorem V.1.2, that $1 - 2/z^2$ also has Julia set $\mathcal{J} = \overline{\mathbf{C}}$.

## 2.  Counting Cycles

In this section we show that the number of attracting and neutral cycles of a rational map is finite. Along the way we show that each basin of attraction for an attracting or parabolic cycle contains a critical point. This bounds the number of such cycles by the number of critical points.

Recall the basin of attraction of an attracting fixed point $z_0$ is the set $A(z_0) = A(z_0, R)$ consisting of $z$ such that $R^n(z) \to z_0$. If $\{z_0, ..., z_{m-1}\}$ is an attracting cycle of length $m$, then each $z_j$ is an attracting fixed point for $R^m$, and we define the *basin of attraction* of the attracting cycle, or of $z_0$, to be the union of the basins of attraction $A(z_j, R^m)$ of the $z_j$'s with respect to $R^m$. The basin of attraction is again denoted by $A(z_0)$. The *immediate basin of attraction* of the cycle, denoted by $A^*(z_0)$, is the union of the $m$ components of $A(z_0)$ which contain points of the cycle.

THEOREM 2.1. *If $z_0$ is an attracting periodic point, then the basin of attraction $A(z_0)$ is a union of components of the Fatou set, and the boundary of $A(z_0)$ coincides with the Julia set.*

*Proof.* Let $U$ be an open neighborhood of the cycle of $z_0$ contained in the Fatou set. Then $A(z_0)$ is the union of the backward iterates of $U$, an open subset of the Fatou set. If $w_0 \in \partial A(z_0)$ and $V$ is any neighborhood of $w_0$, then the iterates $R^n(z)$ converge towards the cycle of $z_0$ on $V \cap A(z_0)$, whereas they remain outside $A(z_0)$ for $z \in V \backslash A(z_0)$. Consequently $\{R^n\}$ is not normal on $V$, and $w_0 \in \mathcal{J}$. Since $A(z_0)$ is completely invariant, Theorem 1.7 gives $\mathcal{J} = \partial A(z_0)$. $\square$

THEOREM 2.2. *If $z_0$ is an attracting periodic point, then the immediate basin of attraction $A^*(z_0)$ contains at least one critical point.*

*Proof.* Suppose first that $z_0$ is an attracting fixed point. We may assume its multiplier $\lambda$ satisfies $0 < |\lambda| < 1$. Let $U_0 = \Delta(z_0, \varepsilon)$ be a small disk, invariant under $R$, on which the analytic branch $f$ of $R^{-1}$ satisfying $f(z_0) = z_0$ is defined. The branch $f$ maps $U_0$ into $A^*(z_0)$, and $f$ is one-to-one. Thus $U_1 = f(U_0)$ is simply connected, and $U_1 \supset U_0$. We proceed in this fashion, constructing $U_{n+1} = f(U_n) \supset U_n$ and extending $f$ analytically to $U_{n+1}$. If the procedure does not terminate we obtain a sequence $f^n : U_0 \to U_n$ of analytic functions on $U_0$ which omits $\mathcal{J}$, hence is normal on $U_0$. But this is impossible, since $z_0 \in U_0$ is a repelling fixed point for $f$. Thus we eventually reach a $U_n$ to which we cannot extend $f$. There is then a critical point $p \in A^*(z_0)$ such that $R(p) \in U_n$.

If $z_0$ is periodic with period $n > 1$ and $|(R^n)'(z_0)| < 1$, this argument shows each component of $A^*(z_0)$ contains a critical point of $R^n$. Since $(R^n)'(z) = \prod R'(R^j(z))$, $A^*(z_0)$ must also contain a critical point of $R$. $\square$

REMARK. We observe for later reference that the critical point $p$ produced above, for an attracting cycle that is not superattracting, has distinct iterates $R^k(p)$, $k \geq 1$. This is because $R$ is one-to-one on $U_n$.

Since there are $2d - 2$ critical points, counting multiplicity, Theorem 2.2 shows that the number of attracting cycles is at most $2d - 2$.

A proof of Theorem 2.2 can be based on Koenigs' coordinate function $\varphi$ and the (locally defined) inverse $\varphi^{-1}$ near 0. Recall that $\varphi$ maps $A^*(z_0)$ onto $\mathbf{C}$, $\varphi(z_0) = 0$, and $\varphi'(z_0) \neq 0$. Since $A^*$ omits three points and hence cannot be mapped conformally onto $\mathbf{C}$, the radius of convergence $\rho$ of the power series for $\varphi^{-1}(\zeta)$ at 0 is finite. There must be a critical point for $R$ on the boundary of $\varphi^{-1}(|\zeta| < \rho)$, or else the functional equation $\varphi^{-1}(\zeta) = R^{-1}(\varphi^{-1}(\lambda\zeta))$ could be used to extend $\varphi^{-1}$ to a larger disk.

Next we deal with the rationally neutral fixed points. First suppose $z_0$ is a fixed point with multiplier 1, and let $P$ be an attracting petal at $z_0$. The *basin of attraction* associated with $P$ consists of all $z$ such that $R^n(z) \in P$ for some $n \geq 0$, that is, such that $R^n(z)$ tends to $z_0$ through $P$. As in the case of attracting cycles, the basin of attraction is an open subset of the Fatou set $\mathcal{F}$, whose boundary coincides with the Julia set. The *immediate basin of attraction* is the connected component of the basin of attraction containing $P$.

The definitions are extended in the obvious way to arbitrary rationally neutral cycles, as follows. Suppose $\{z_0, ..., z_{n-1}\}$ is a parabolic cycle of length $n$, so that the multiplier of $R^n$ at the fixed point $z_0$ is a primitive $m$th root of unity. Let $P$ be an attracting petal for $R^{nm}$ at $z_0$. We define the *basin of attraction* of $R$ containing the petal $P$ to consist of all $z$ such that $R^k(z) \in P$ for some $k \geq 1$. In this case, the sets $R^j(P)$, $0 \leq j < nm$, are disjoint attracting petals at the $z_j$'s, and the basin of attraction of $R$ containing $P$ is the union of the basins of attraction of $R^{nm}$ containing the $R^j(P)$'s. The *immediate basin of attraction* containing the petal $P$ is the union of the $nm$ connected components of $\mathcal{F}$ containing the $R^j(P)$'s.

THEOREM 2.3. *If $z_0$ is a rationally neutral periodic point, then each immediate basin of attraction associated with the cycle of $z_0$ contains a critical point.*

*Proof.* Replacing $R$ by $R^N$, we may assume $R(z_0) = z_0$ and $R'(z_0) = 1$. Let $A^*$ be the immediate basin of attraction containing a petal $P$ at $z_0$. Let $\varphi$ be the Fatou coordinate function defined on $P$, so that $\varphi$ is univalent and conjugates $R$ to a translation: $\varphi(R(z)) = \varphi(z)+1$. Since $R(A^*) = A^*$, the functional equation allows us to extend $\varphi$ to $A^*$, where it satisfies the same identity. Since $\varphi(P)$ covers a right half-plane, $\varphi$ maps $A^*$ onto the entire complex plane. Now $\varphi$ has a critical point $w_0$ in $A^*$, else we could define a branch of $\varphi^{-1}$, which would be a meromorphic function on the entire complex plane omitting the Julia set, hence constant. For $n$ large, $R^n(w_0)$ is in $P$, where $\varphi$ has no critical points, so by replacing $w_0$ by the largest iterate that is a critical point, we can assume that $R(w_0)$ is not a critical point of

$\varphi$. From the functional equation we then have $\varphi'(R(w_0))R'(w_0) = \varphi'(w_0) = 0$, so that $w_0$ is a critical point of $R$. $\square$

THEOREM 2.4. *The number of attracting cycles plus the number of rationally neutral cycles (counting their natural multiplicity) is at most $2d - 2$.*

*Proof.* This follows immediately from Theorems 2.2 and 2.3, since the basins of attraction are disjoint, and since there are at most $2d - 2$ critical points. $\square$

Finally we treat irrationally neutral cycles by means of an analytic perturbation.

LEMMA 2.5. *Let $R_t(z)$ be a family of rational functions depending analytically on a parameter $t$. Let $z_0$ be a periodic point of $R_{t_0}$ of period $m$, and suppose the multiplier of the cycle of $z_0$ is $\neq 1$. Then for $t$ near $t_0$ there is a unique periodic point $z(t)$ of $R_t$ of period $m$ near $z_0$. The periodic point $z(t)$ depends analytically on $t$, and $z(t_0) = z_0$. The multiplier $\lambda(t)$ of the cycle is analytic, and $z(t)$ and $\lambda(t)$ extend analytically along any path in the $t$-plane along which $R_t$ is defined provided $\lambda(t)$ remains bounded away from 1, so that $z(t)$ is periodic of period $m$.*

*Proof.* Set $Q(z,t) = R_t^m(z) - z$, and let $(R_t^m)'(z)$ be the $z$-derivative of $R_t^m(z)$. A point $z$ is periodic for $R_t$ with period dividing $m$ if and only if $Q(z,t) = 0$. If the period is exactly $m$, then the multiplier of the cycle is $\lambda(t) = (R_t^m)'(z)$. Now the $z$-derivative of $Q(z,c)$ is $(R_t^m)'(z) - 1$. Thus the implicit function theorem guarantees that if the multiplier of the cycle of $z_0$ is not 1, the equation $Q(z,t) = 0$ has a unique solution $z(t)$ near $z_0$ for $t$ near $t_0$, and $z(t)$ depends analytically on $t$, as does the multiplier $\lambda(t)$ of the cycle of $z(t)$. The period of each $z(t)$ divides $m$, and the set of $t$-values for which it is equal to $m$ is evidently open, so the period must remain constant along any path. $\square$

THEOREM 2.6. *The number of attracting cycles plus half the number of neutral cycles with multiplier $\neq 1$ is at most $2d - 2$.*

*Proof.* We consider an analytic perturbation $R_t$ of $R = R_0$. Under an analytic perturbation we can expect half the neutral cycles to become attracting. To foreclose the possibility that the multiplier of the perturbed cycle is constant, we choose our perturbation to be an analytic deformation of $R$ to $z^d$, which has no neutral cycles. If $R = P/Q$, then such a deformation is given for instance by $R_t(z) = [(1 - t)P(z) + tz^d]/[(1 - t)Q(z) + t]$, defined for all complex $t$ for which $R_t$ has degree $d$ (a cofinite subset of $\mathbf{C}$).

Let $z_0$ be a neutral periodic point of $R$, with period $m$ and with multiplier $\lambda_0 \neq 1$. Let $z(t)$ be as in the preceding lemma, with multiplier $\lambda(t)$. Since we can continue $z(t)$ and $\lambda(t)$ analytically along some path to 1, and $|\lambda(1)| \neq 1$, $\lambda(t)$ is not constant. Let $E_\rho$ be the set of $\theta$ such that $|\lambda(\rho e^{i\theta})| < 1$. Then the length of $E_\rho$ tends to $\pi$ as $\rho \to 0$.

Define these sets for one point in each neutral cycle. Suppose there are $N$ such cycles. Let $E_{j,\rho}$ be the corresponding set and let $\chi_{j,\rho}(\theta)$ be the corresponding characteristic function on the circle. Then

$$\sum_{j=1}^{N} \frac{1}{2\pi} \int_0^{2\pi} \chi_{j,\rho}(\theta)d\theta \geq \frac{1}{2}N - \varepsilon(\rho),$$

where $\varepsilon(\rho) \to 0$ as $\rho \to 0$. Hence for $\rho$ small, there is a $\theta$ with $\sum \chi_{j,\rho}(\theta) \geq N/2$. If $\rho$ is small enough, then the old attracting cycles of $R$ still correspond to attracting cycles for $R_t$ and at least half of the neutral cycles do also. By Theorem 2.4, $R_t$ has at most $2d - 2$ attracting cycles and this completes the proof. $\square$

Theorems 2.4 and 2.6 combined yield immediately the finiteness theorem for nonrepelling cycles, with the estimate $6d - 6$. The proof method we have followed is due to Fatou.

THEOREM 2.7. *The total number of attracting and neutral cycles is at most $6d - 6$.*

The sharp value for Theorem 2.7 is $2d - 2$, due to M. Shishikura (1987), who obtained the result for his Master's thesis. He showed, using quasiconformal surgery, that there is an analytic perturbation for which all neutral cycles become attracting (see [**Be2**]). We prove this later for polynomials (Theorem VI.1.2), in which case the sharp estimate of $d - 1$ bounded nonrepelling cycles is due to A. Douady (1982).

# 3.  Density of Repelling Periodic Points

We can now prove the theorem, obtained independently by Julia and Fatou, which was mentioned at the beginning of the chapter.

**THEOREM 3.1.** *The Julia set $\mathcal{J}$ is the closure of the repelling periodic points.*

*Proof.* Suppose there is an open disk $U$ that meets $\mathcal{J}$ and that contains no fixed points of any $R^m$. We may assume $U$ contains no poles of $R$ nor critical values of $R$. If $f_1, f_2$ are two different branches of $R^{-1}$ on $U$, then since there are no solutions of $R^m(z) = z$ in $U$,

$$g_n = \frac{R^n - f_1}{R^n - f_2} \cdot \frac{z - f_2}{z - f_1}$$

omits the values $0, 1, \infty$ in $U$. By Montel's theorem, $\{g_n\}$ is normal and hence so is $\{R^n\}$, a contradiction. Thus periodic points are dense in $\mathcal{J}$. Since there are only a finite number of attracting and neutral cycles, and since $\mathcal{J}$ is perfect, the repelling cycles are dense. $\square$

**THEOREM 3.2.** *Let $U$ be open, and suppose $U \cap \mathcal{J} \neq \emptyset$. Then there is an integer $N$ such that $R^N(U \cap \mathcal{J}) = \mathcal{J}$.*

*Proof.* Let $z_0 \in U \cap \mathcal{J}$ be a repelling periodic point with period $n$. Choose $V$ open so that $z_0 \in V \subset U$ and $V \subset R^n(V)$. Then $R^{n(j-1)}(V) \subset R^{nj}(V)$, and by Theorem 1.5

$$\mathcal{J} \subset \overline{\mathbf{C}} \backslash E \subset \cup_{j=1}^\infty R^{nj}(V).$$

Since $\mathcal{J}$ is compact, the Heine-Borel theorem implies $\mathcal{J} \subset R^{jn}(V)$ for some $j$. Thus the theorem holds with $N = nj$. $\square$

Theorem 3.2 shows that $\mathcal{J}$ is "almost everywhere" conformally self-similar. More explicitly, if $z_0 \in \mathcal{J}$ has no critical point in its forward orbit, then within any neighborhood of any other point of $\mathcal{J}$ there is an open subset of $\mathcal{J}$ that is the conformal image of a neighborhood of $z_0$.

An open subset $A$ of $\overline{\mathbf{C}}$ is a *conformal annulus* if $A$ can be mapped conformally to a genuine annulus $\{r < |\zeta| < s\}$. The *module* of $A$ is defined to be $\log(s/r)$ times a normalization factor $1/2\pi$, and this is a conformal invariant. A compact subset $E$ of $\overline{\mathbf{C}}$ is *uniformly perfect* if $E$ has at least two points and if the modules of the conformal annuli in $\overline{\mathbf{C}}\backslash E$ which separate $E$ are bounded. As an application of the Fatou–Julia theorem, we prove the following theorem of R. Mañé and L.F. da Rocha (1992).

THEOREM 3.3. *The Julia set $\mathcal{J}$ of $R$ is uniformly perfect.*

*Proof.* Suppose there is a sequence $\{A_n\}$ of conformal annuli in $\mathcal{F}$ with modules tending to $\infty$, such that both components of $\overline{\mathbf{C}}\backslash A_n$ meet $\mathcal{J}$. Let $E_n$ be the component of $\overline{\mathbf{C}}\backslash A_n$ with the smaller diameter (measured in the spherical metric). The diameter of $E_n$ tends to 0, as can be seen by noting that there are closed geodesics (conformal circles) in $A_n$ whose hyperbolic lengths tend to 0 as $n \to \infty$, and by using the usual comparison of metrics. Let $\psi_n$ map the open unit disk conformally onto $A_n \cup E_n$, so that $\psi_n(0) \in E_n$. Then $K_n = \psi_n^{-1}(E_n)$ is a continuum in $\Delta$ containing 0. The module of $\Delta\backslash K_n$ also tends to $\infty$, since it is the same as that of $A_n$. Hence the diameter of $K_n$ tends to 0, as above.

Fix $\delta > 0$ small, and assume that the diameter of $E_n$ is less than $\delta$. In view of Theorem 3.2, there is a first integer $k_n$ such that the diameter of $R^{k_n}(E_n)$ exceeds $\delta$. If $g_n = R^{k_n} \circ \psi_n$, then the diameter of $g_n(K_n)$ is at least $\delta$. Now the diameter of $R^{k_n}(E_n)$ is at most $C\delta$, where $C$ is the Lipschitz constant of $R$ (with respect to the spherical metric). Fix four points of $\mathcal{J}$ which are of distance greater than $C\delta$ from each other. Since $g_n(\Delta\backslash K_n) \subset \mathcal{F}$ and $g_n(K_n)$ has diameter at most $C\delta$, each $g_n$ omits at least three of the four (fixed) points. Consequently $\{g_n\}$ is a normal family, and the diameter of $g_n(K_n)$ tends to 0. This contradiction establishes the theorem. $\square$

A simple scaling and normal families argument shows that conformal annuli of large module contain genuine annuli of large module. Thus the compact set $E$ is uniformly perfect if and only if there is $c > 0$ such that for any finite $z_0 \in E$ and $r > 0$ (and $r < r_0$ when $\infty \notin E$), the euclidean annulus $\{cr < |z - z_0| < r\}$ meets $E$. Uniformly perfect sets were introduced by A.F. Beardon and Ch. Pommerenke (1979), who showed that if $\infty \notin E$, the compact set $E$ is uniformly perfect if and only if the hyperbolic metric of the complement of $E$ is comparable to the reciprocal of the distance to the boundary, that is, the factor $\log(1/\delta(z))$ appearing in Theorem I.4.3

can be omitted. There are many other characterizations of uniformly
perfect sets. One characterization is in terms of linear density of logarithmic capacity (see [**Po2**]). This guarantees that Green's function
of any fixed component of the Fatou set extends to the boundary to
be Hölder continuous, and this in turn implies that Julia sets have
positive Hausdorff dimension. For more details, see Section VIII.3,
where Hölder continuity is proved in the context of quadratic polynomials.

## 4.   Polynomials

In this section we consider the case where $R = P$ is a polynomial
of degree $d \geq 2$. We have a superattracting fixed point at $\infty$. The
iterates of $P$ are bounded on the bounded components of $\mathcal{F}$, by the
maximum principle, so the basin of attraction $A(\infty)$ is connected.
By Theorem III.1.7, the Julia set coincides with $\partial A(\infty)$.

The *filled-in Julia set* of $P$, denoted by $\mathcal{K}$, is defined to be the
union of the Julia set $\mathcal{J}$ and the bounded components of $\mathcal{F}$. Thus
$z \in \mathcal{K}$ if and only if the iterates $P^n(z)$ are bounded. This property
can be taken as a basis for drawing computer pictures of $\mathcal{K}$ and of
$\mathcal{J}$. For $\mathcal{K}$, color $z$ purple if $|P^n(z)| < C$ for $1 \leq n \leq N$, otherwise
color $z$ white. For $\mathcal{J}$ simply recolor $z$ white if it and all its neighbors
are purple. This algorithm for computing $\mathcal{J}$ is called the *boundary
scanning method*.

Let $\varphi$ conjugate $P(z)$ to $\zeta^d$ near $\infty$, with $\varphi(z) = z + \mathcal{O}(1)$ at $\infty$. As
discussed in Section II.4, $\log|\varphi(z)|$ coincides with Green's function
$G(z)$ for $A(\infty)$ with pole at $\infty$. The functional equation for $\varphi(z)$
gives a functional equation for Green's function,

$$G(P(z)) = d \cdot G(z), \qquad z \in A(\infty).$$

Thus $P$ maps level curves of $G$ to level curves, increasing $d$-fold
the value of Green's function. Green's function provides a precise
measure of the escape rate to $\infty$. The exterior $\{G > r\}$ of the level
curve is invariant under $P$, and $P$ maps it $d$-to-one onto $\{G > rd\}$.
For $r$ large, $\varphi(z)$ is defined on $\{G > r\}$ and maps it conformally onto
$\{|\zeta| > e^r\}$. The equation $\varphi(z) = (\varphi(P(z)))^{1/d}$ allows us to extend
$\varphi(z)$ to $\{G > r/d\}$ provided no critical point of $P$ belongs to this
domain.

Now there are two cases to consider. If there is no critical point

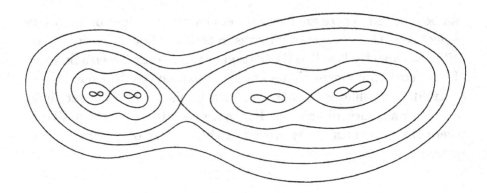

FIGURE 1. Level lines of Green's function with $\mathcal{J}$ totally disconnected.

of $P$ in $A(\infty)$, we can continue $\varphi$ indefinitely to all of $A(\infty)$, and $\varphi$ maps $A(\infty)$ conformally onto the complement $\{|\zeta| > 1\}$ of the closed unit disk in the $\zeta$-plane. In particular, $A(\infty)$ is simply connected, and the Julia set $\mathcal{J} = \partial A(\infty)$ is connected.

Otherwise we extend $\varphi$ until we reach a level line $\{G = r\}$ of Green's function that contains a critical point of $P$. The situation is then as follows. The domain $\{G > r\}$ is simply connected and mapped by $\varphi$ conformally onto $\{|\zeta| > e^r\}$. The domain forms several cusps at the critical point, and $\varphi(z)$ approaches different values as $z$ approaches the critical point through different cusps. The level line $\{G = r\}$ consists of at least two simple closed curves that meet at the critical point. Within each of these curves there are points of $\mathcal{J}$, or else $G$ would be harmonic and positive, hence constant within the curve. Hence $\mathcal{J}$ is disconnected. In fact, in this case $\mathcal{J}$ has uncountably many connected components. This can be seen by noting that the critical points of $G$ are the critical points of $P$ and all their inverse iterates, and by following the splitting of level curves at each such critical point.

We have proved in particular the following.

THEOREM 4.1. *The Julia set $\mathcal{J}$ is connected if and only if there is no finite critical point of $P$ in $A(\infty)$, that is, if and only if the forward orbit of each finite critical point is bounded.*

At the other extreme we have the following.

THEOREM 4.2. *If $P^n(q) \to \infty$ for each critical point $q$, then the Julia set $\mathcal{J}$ is totally disconnected.*

*Proof.* Let $D$ be a large open disk containing $\mathcal{J}$ such that $P(\overline{\mathbb{C}}\backslash D) \subset \overline{\mathbb{C}}\backslash\overline{D}$. Choose $N$ so large that $P^N$ maps the critical points of $P$ to $\overline{\mathbb{C}}\backslash D$. Since the critical points of $P^n$ are the critical points of $P$ and their first $n-1$ iterates, there are no critical points of $P^n$ in $P^{-n}(\overline{D})$ for $n \geq N$. Thus for $n \geq N$ there are no critical values of $P^n$ in $\overline{D}$, so that all inverse branches $P^{-n}$ are defined and map $\overline{D}$ conformally into $D$. Let $z_0 \in \mathcal{J}$. Then $P^n(z_0) \in \mathcal{J}$, and we define $f_n$ to be the inverse branch of $P^n$ which maps $P^n(z_0)$ to $z_0$. The $f_n$'s are uniformly bounded on a neighborhood of $\overline{D}$, hence they form a normal family there. Since $f_n(z)$ accumulates on $\mathcal{J}$ for $z \in D \cap A(\infty)$, any limit function $f$ maps $D \cap A(\infty)$ into $\mathcal{J}$. Since $\mathcal{J}$ contains no open sets (Theorem 1.9), $f$ must be constant. Hence $f_n(\overline{D})$ has diameter tending to zero. Since $f_n(\partial D)$ is disjoint from $\mathcal{J}$, $\{z_0\}$ must be a connected component of $\mathcal{J}$, and $\mathcal{J}$ is totally disconnected. $\square$

If $P$ is quadratic, then Theorems 4.1 and 4.2 cover all cases, and the corresponding Julia set is either connected or totally disconnected. We study quadratic polynomials in detail in Chapter VIII, in connection with the Mandelbrot set.

# IV

# Classification of Periodic Components

We focus on the behavior of a rational function $R(z)$ on the Fatou set $\mathcal{F}$. Our aim is twofold: to show that every component of $\mathcal{F}$ is iterated eventually to a periodic component, and to classify the action of $R(z)$ on periodic components.

## 1. Sullivan's Theorem

As before, we assume that the degree of $R$ is $d \geq 2$. The image of any component of the Fatou set $\mathcal{F}$ under $R$ is a component of $\mathcal{F}$, and the inverse image of a component of $\mathcal{F}$ is the disjoint union of at most $d$ components of $\mathcal{F}$. The dynamics of $R$ can be understood in part by determining how the various components of $\mathcal{F}$ are moved about by $R$.

Consider a fixed component $U$ of $\mathcal{F}$. There are several possibilities for the orbit of $U$ under $R$.

1. If $R(U) = U$, we call $U$ a *fixed component* of $\mathcal{F}$.

2. If $R^n(U) = U$ for some $n \geq 1$, we call $U$ a *periodic component* of $\mathcal{F}$. The minimal $n$ is the *period* of the component. If $n = 1$, we have a fixed component.

3. If $R^m(U)$ is periodic for some $m \geq 1$, we call $U$ a *preperiodic component* of $\mathcal{F}$.

4. Otherwise, all $\{R^n(U)\}$ are distinct, and we call $U$ a *wandering domain*.

We have already seen several examples of fixed and periodic components of $\mathcal{F}$, and we will soon see strictly preperiodic components. I.N. Baker (1976) has shown that some entire functions have wandering domains. However, according to D. Sullivan (1985) this is not possible for rational functions, and our main goal in this section is to prove this. First we make some preliminary observations.

On any component $U$ of $\mathcal{F}$, $R$ is a branched cover of $U$ over $R(U)$ with at most $d$ sheets. A component of $\mathcal{F}$ is completely invariant if and only if $R$ is a $d$-sheeted branched covering of the component over itself.

THEOREM 1.1. *If $U$ is a completely invariant component of $\mathcal{F}$, then $\partial U = \mathcal{J}$, and every other component of $\mathcal{F}$ is simply connected. There are at most two completely invariant components of $\mathcal{F}$.*

*Proof.* If $U$ is completely invariant, then $\partial U = \mathcal{J}$, by Theorem III.1.7. Moreover the sequence $\{R^n\}$ omits the open set $U$ on $\overline{\mathbf{C}}\backslash\overline{U}$, so $\{R^n\}$ is normal there, and $\overline{\mathbf{C}}\backslash\overline{U} \subset \mathcal{F}$. Since $U$ is connected, each component of $\overline{\mathbf{C}}\backslash\overline{U}$ is simply connected. If $U$ is furthermore simply connected, then since $R$ is a $d$-to-1 mapping, $U$ must contain $d-1$ critical points. Since there are only $2d-2$ critical points altogether, there can be at most two simply connected completely invariant components. $\square$

THEOREM 1.2. *The number of components of the Fatou set can be $0$, $1$, $2$, or $\infty$, and all cases occur.*

*Proof.* Suppose $\mathcal{F}$ has only finitely many components and let $U_0$ be one. Consider a chain of inverse images $R(U_{-1}) = U_0$, $R(U_{-2}) = U_{-1}, \cdots$. Eventually we reach $U_{-n}$ with $R(U_{-n}) = U_{-k}$ for some $0 \leq k < n$. Then $U_0 = R^n(U_{-n}) = R^n(U_{-k}) = R^{n-k}(U_0)$. Thus each component $U_0$ of $\mathcal{F}$ is periodic, and since there are only finitely many, there is some $N$ such that $R^N(U) = U$ for every component $U$. Hence every component is completely invariant for $R^N$, and by the preceding theorem there are at most two components.

We have seen in the previous chapter that the map $R(z) = z^2$ has two components in its Fatou set and $R(z) = z^2 - 2$ has one. The Lattès example in Section III.1 has empty Fatou set $\mathcal{F}$ and Julia set $\mathcal{J} = \overline{\mathbf{C}}$. Finally, if $P(z) = \lambda z + z^2$ is chosen to correspond to a Siegel disk, then $\mathcal{F}$ has infinitely many components. To see this, let $U_0$ and $U_\infty$ be the components containing 0 and $\infty$ respectively. Since $R$ is conjugate to a rotation of $U_0$ there are no critical points in $U_0$ and $R$ is one-to-one on $U_0$. Thus $U_0$ has a distinct preimage $U_1$, which is also distinct from $U_\infty$, so $\mathcal{F}$ has infinitely many components. $\square$

THEOREM 1.3 (Sullivan). *A rational map has no wandering domains.*

*Proof.* Assume $U_0$ is wandering and let $U_n = R^n(U_0), n \geq 1$. We assume $\infty$ is in some component $V$ of $\mathcal{F}$ other than the $U_n$'s. This implies that $\sum \text{area} \, U_n < \infty$, which implies that the only possible limit functions of $\{R^n\}$ on $U_0$ are constants. In fact, $(R^n)' \to 0$ on compact subsets for otherwise each $U_n$ would contain a disk with diameter bounded uniformly away from 0. Replacing $U_n$ by $U_{n+m}$, we may assume no $U_n$ contains a critical point of $R$.

We claim that $R$ maps each $U_n$ one-to-one onto $U_{n+1}$. For this it suffices to prove that each $\{U_n\}$ is simply connected. Let $\gamma$ be a closed curve in $U_0$. Since $(R^n)' \to 0$ on $U_0$, the diameters of the images $R^n(\gamma)$ tend to 0. Now each pole of $R$ is contained in a component of $\mathcal{F}$ that is mapped to $V$, and the images of $\gamma$ are not mapped to $V$, so that if the diameter of $R^n(\gamma)$ is sufficiently small, say for $n \geq N$, there are no poles of $R$ in the bounded components of $\mathbf{C} \backslash R^n(\gamma)$. Then the iterates $R^{N+k}$ are all analytic inside $\mathbf{C} \backslash R^N(\gamma)$, and by the maximum modulus principle they are uniformly bounded there, so that the bounded components of $\mathbf{C} \backslash R^N(\gamma)$ are included in $\mathcal{F}$, hence in $U_N$. Then $R^N(\gamma)$ is homotopic to a constant path in $U_N$. Since there are no critical points, $R^N$ is a covering map of $U_0$ over $U_N$, and the homotopy can be lifted to a homotopy of $\gamma$ to a constant in $U_0$. Hence $U_0$ is simply connected, as is each $U_n$ by the same argument.

We will now construct a family of quasiconformal mappings $\{f_t\}$ with $t \in \mathbf{C}^N$, so that $f_t^{-1} \circ R \circ f_t$ is an $N$-dimensional analytic family of distinct rational maps. Since they have the same degree as $R$, this gives a contradiction if we take $N > 2d - 1$.

So fix $N > 2d - 1$, and suppose $D_0 = \Delta(0, \varepsilon) \subset D = \Delta(0, 2\varepsilon) \subset U_0$. For $t = (t_1, ..., t_N)$ in $\mathbf{C}^N$, $|t| < \delta$, we define a Beltrami coefficient (ellipse field) $\mu$ on $D_0$ by

$$\mu(z) = \sum_{1}^{N} t_j e^{-ij\theta}, \qquad z = re^{i\theta} \in D_0.$$

For $n \geq 1$, set $D_n = R^n(D_0)$. Since the $D_n$'s are disjoint and $R$ maps each $D_n$ conformally onto $D_{n+1}$, we can extend the ellipse field to $\cup D_n$ to be invariant under $R$. Each component of $R^{-1}(\cup D_n)$ is mapped finite-to-one onto one of the $D_n$'s, so we can further extend the ellipse field to $R^{-1}(\cup D_n)$, and proceeding backwards to all inverse images of $\cup D_n$ to be invariant under $R$. Since $R$ is conformal, the extension does not increase $\|\mu\|_\infty$. For other $z$ set $\mu = 0$, and then the ellipse field is everywhere invariant. Note that $\mu = 0$ on $U_0 \backslash D_0$.

Let $f_t(z) = f(z,t)$ be the solution of the Beltrami equation, normalized so $f(z,t) = z + o(1)$ at $\infty$. Now (suppressing the $t$'s) let $g = f_z - 1$ and recall from Section I.7 that

$$g = (I - U_\mu)^{-1} S(\mu) = S(\mu) + U_\mu(S(\mu)) + U_\mu^2(S(\mu)) + \cdots .$$

Since $\|\mu\|_\infty = \mathcal{O}(|t|)$, also $\|S(\mu)\|_2 = \mathcal{O}(|t|)$, $\|U_\mu\|_2 = \mathcal{O}(|t|)$, and consequently $g = S(\mu) + \mathcal{O}(|t|^2)$ in $L^2$. Thus with $\zeta = \xi + i\eta = \rho e^{i\phi}$, we calculate for $z \in D \backslash D_0$ that

$$\left. \frac{\partial g}{\partial t_j} \right|_{t=0} = \left. \frac{\partial S(\mu)}{\partial t_j} \right|_{t=0} = -\frac{1}{\pi} \iint_{D_0} \frac{e^{-ij\phi}}{(\zeta - z)^2} d\xi d\eta + \varphi_j(z) = \frac{c_j}{z^{j+2}} + \varphi_j(z),$$

where the $\varphi_j$'s are analytic in $D$, and $c_j = -2\varepsilon^{j+2}(j+1)/(j+2) \neq 0$. It follows that if $\lambda = (\lambda_j) \in \mathbf{C}^N$, $\lambda \neq 0$, then for some $z \in D \backslash D_0$

$$\sum \lambda_j \frac{\partial g}{\partial t_j}(z,0) \neq 0, \tag{1.1}$$

because otherwise

$$\sum \frac{\lambda_j c_j}{z^{j+2}} \equiv -\sum \lambda_j \varphi_j, \qquad \varepsilon < |z| < 2\varepsilon,$$

which contradicts the uniqueness of Laurent expansions.

From the invariance of the ellipse field, we see that $R_t = f_t \circ R \circ (f_t)^{-1}$ is analytic. (See the remark at the end of Chapter I.) It is furthermore $d$-to-1. Thus it is a rational mapping of the same degree as $R$. By Theorem I.7.6, $f_t$ depends analytically on the parameter $t$, and moreover $f_0$ is the identity $z$. Hence $R_t$ depends analytically on the parameter $t$, and $R_0 = R$. Thus the (normalized) coefficients of $R_t$ are holomorphic functions of $t$ and agree with those of $R$ at $t = 0$. Write, assuming $R(\infty) = 1$,

$$R_t(z) = \frac{a_d(t)z^d + a_{d-1}z^{d-1} + \cdots + a_0(t)}{z^d + b_{d-1}(t)z^{d-1} \cdots + b_0(t)},$$

where $a_d(0) = 1$ and $a_i(t), b_i(t)$ are holomorphic in $\{|t| < \delta\}$. Let $V$ be the connected component containing 0 of the analytic subvariety of $\{|t| < \delta\}$ determined by the equations $a_i(t) = a_i(0), b_i(t) = b_i(0)$. There are only $2d-1$ equations here, and so $V$ has positive dimension. (For background on analytic varieties, see [GuR].)

Let $\tau \in V$. Then $f_\tau \circ R_0 \circ (f_\tau)^{-1} = R$. If $z_0$ is a fixed point of $R^n$, then so is $f_\tau(z_0)$, and vice versa, since $f_\tau(z_0) = f_\tau(R^n(z_0)) = R^n(f_\tau(z_0))$. If $z_0$ is a repelling fixed point for $R^n$ then so is $f_\tau(z_0)$. Hence the set $A_n$ of such fixed points satisfies $f_\tau(A_n) = A_n$. However, since $f_0$ is the identity and $A_n$ is discrete, $f_\tau$ must fix each point of $A_n$. But $\cup A_n$ is dense in $\mathcal{J}$, by Theorem II.2.8, and so $f_\tau(z) = z$ on $\mathcal{J}$. Now $f_\tau(z)$ is analytic on $U_0 \backslash D_0$ and $f_\tau(z) = z$ on $\partial U_0 \subseteq \mathcal{J}$. It follows that $f_\tau(z) = z$ in $U_0 \backslash D_0$. (For an elementary argument, map $U_0$ conformally to the unit disk and observe that $f_\tau(z) - z$ becomes analytic on an annulus $\{r < |\zeta| < 1\}$ and tends to 0 as $|\zeta| \to 1$, so is identically 0 on the annulus.)

Now changing notation, we have $f(z, \tau) - z = 0$ for all $z \in U_0 \backslash D_0$, $\tau \in V$. Since $g = f_z - 1$, we obtain

$$g(z, \tau) = 0, \qquad z \in U_0 \backslash D_0, \tau \in V. \qquad (1.2)$$

If $\tau_0$ is a regular point of $V$, there is an analytic map $\zeta \to \tau(\zeta)$ of a disk $\Delta(0, \delta)$ into $V$ with $\tau(0) = \tau_0$ and $\tau'(0) = \lambda$, $|\lambda| = 1$. From (1.2) and the chain rule we obtain

$$\sum \lambda_j \frac{\partial g}{\partial t_j}(z, \tau_0) = 0, \qquad z \in U_0 \backslash D_0.$$

Here the unit vector $\lambda$ depends on $\tau_0$, and passing to a limit of such $\lambda$'s as $\tau_0 \to 0$ we obtain a contradiction to (1.1). We conclude there are no wandering domains. $\square$

## 2.   The Classification Theorem

Having ruled out the possibility of wandering domains, we now know that every component of the Fatou set is periodic or preperiodic. Our next goal is to classify periodic components. We have already seen some examples: domains containing an attracting fixed point, and Siegel disks. There are various other possibilities.

A periodic component $U$ of period $n$ of the Fatou set $\mathcal{F}$ is called *parabolic* if there is on its boundary a neutral fixed point $\zeta$ for $R^n$ with multiplier 1, such that all points in $U$ converge to $\zeta$ under iteration by $R^n$. The domains $U, R(U), ..., R^{n-1}(U)$ form a *parabolic cycle*. Their union is the immediate basin of attraction associated with an attracting petal at $\zeta$.

A periodic component $U$ of period $n$ of the Fatou set is called a *Herman ring* (or an *Arnold ring*) if it is doubly connected and $R^n$ is conjugate to either a rotation on an annulus or to a rotation followed by an inversion. We shall see in Chapter VI that Herman rings can occur and that moreover there can be only finitely many of them. The definition of Siegel disk is similarly extended to include periodic components $U$ that are simply connected, on which some $R^n$ is conjugate to a rotation. Siegel disks and Herman rings are referred to as *rotation domains*.

THEOREM 2.1. *Suppose $U$ is a periodic component of the Fatou set $\mathcal{F}$. Then exactly one of the following holds:*

1. *$U$ contains an attracting periodic point.*

2. *$U$ is parabolic.*

3. *$U$ is a Siegel disk.*

4. *$U$ is a Herman ring.*

*Proof.* We may assume $U$ is fixed by $R$. Since $\mathcal{J}$ has more than two points, $U$ is hyperbolic. The proof will be organized as a series of lemmas in the context of analytic maps of hyperbolic domains. Let $\rho = \rho_U$ denote the hyperbolic metric on $U$. The first alternative of Theorem 2.1 is covered by the following lemma.

LEMMA 2.2. *Suppose $U$ is hyperbolic, $f : U \to U$ is analytic, and $f$ is not an isometry with respect to the hyperbolic metric. Then either*

$f^n(z) \to \partial U$ for all $z \in U$, or else there is an attracting fixed point for $f$ in $U$ to which all orbits converge.

*Proof.* Since $f$ is not an isometry, $\rho(f(z), f(w)) < \rho(z, w)$ for all $z, w \in U$. In particular for any compact set $K \subset U$, there is a constant $k = k(K) < 1$ such that

$$\rho(f(z), f(w)) \le k\rho(z, w), \qquad z, w \in K.$$

Suppose there is $z_0 \in U$ whose iterates $z_n = f^n(z_0)$ visit some compact subset $L$ of $U$ infinitely often. Take $K$ to be a compact neighborhood of $L \cup f(L)$. Then $\rho(z_{m+2}, z_{m+1}) \le k\rho(z_{m+1}, z_m)$ whenever $z_m \in L$, and this occurs infinitely often, so $\rho(z_{n+1}, z_n) \to 0$. Thus by continuity any cluster point $\zeta \in L$ of the sequence $\{z_n\}$ is fixed by $f$, and in fact is an attracting fixed point since $\rho(f(z), \zeta) \le k\rho(z, \zeta)$ in some neighborhood of $\zeta$. Since the iterates of $f$ form a normal family, they converge on $U$ to $\zeta$. $\square$

The third and fourth alternatives of Theorem 2.1 are covered by the following.

LEMMA 2.3. *Suppose $U$ is hyperbolic, $f : U \to U$ is analytic, and $f$ is an isometry with respect to the hyperbolic metric. Then exactly one of the following holds:*

1. $f^n(z) \to \partial U$ *for all $z \in U$.*

2. $f^m(z) = z$ *for all $z \in U$ and some fixed $m \ge 1$.*

3. $U$ *is conformally a disk, and $f$ is conjugate to an irrational rotation.*

4. $U$ *is conformally an annulus, and $f$ is conjugate to an irrational rotation or to a reflection followed by an irrational rotation.*

5. $U$ *is conformally a punctured disk, and $f$ is conjugate to an irrational rotation.*

*Proof.* By an isometry, we mean at the local level, so that the lift of $f$ to the universal covering space $\Delta$ is a hyperbolic isometry.

Suppose first that $U$ is simply connected. Let $\varphi$ map $U$ conformally onto the open unit disk $\Delta$. Then $S = \varphi \circ f \circ \varphi^{-1}$ is a conformal self-map of $\Delta$, a Möbius transformation. If $S$ has fixed points on the unit circle, then $|S^n| \to 1$ on $\Delta$, and (1) holds. If $S$ has a fixed point in the disk, we may assume it is at the origin. Then $S$ is a rotation, and either (2) or (3) holds.

Now assume $U$ is not simply connected. Let $\psi : \Delta \to U$ be the universal covering map, and let $\mathcal{G}$ be the associated group of covering transformations, the group of conformal self-maps $g$ of $\Delta$ satisfying $\psi \circ g = \psi$. The lift of $f$ to the unit disk via $\psi$ is a Möbius transformation $F$, which satisfies $\psi \circ F = f \circ \psi$. Let $\Gamma$ be the group obtained by adjoining $F$ to $\mathcal{G}$.

Assume first $\Gamma$ discrete (orbits accumulate only on $\partial \Delta$). Since no iterate $f^k$ of $f$ is the identity on $U$, no iterate $F^k$ of $F$ belongs to $\mathcal{G}$. Since $\Gamma$ is discrete, this implies $gF^k(0) \to \partial\Delta$ uniformly in $g \in \mathcal{G}$. Hence $f^k(z_0) \to \partial U$, and (1) holds.

The final possibility is that $\Gamma$ is not discrete. Let $\overline{\Gamma}$ be the closure of $\Gamma$ in the (Lie) group of conformal self-maps of $\Delta$, and let $\Gamma_0$ be the connected component of $\overline{\Gamma}$ containing the identity. If $g \in \mathcal{G}$ then also $FgF^{-1} \in \mathcal{G}$, since

$$\psi \circ (F \circ g \circ F^{-1}) = f \circ \psi \circ g \circ F^{-1} = f \circ \psi \circ F^{-1} = \psi \circ F \circ F^{-1} = \psi.$$

It follows that $\overline{\Gamma}$, and hence $\Gamma_0$, also conjugates $\mathcal{G}$ to itself. Since $\mathcal{G}$ is discrete and $\Gamma_0$ is connected, $hgh^{-1} = g$ for all $h \in \Gamma_0$ and $g \in \mathcal{G}$, and every $g \in \mathcal{G}$ commutes with every $h \in \Gamma_0$.

LEMMA. *If $A$ and $B$ are two conformal self-maps of the open unit disk $\Delta$ which commute, and $A$ is not the identity, then $B$ belongs to the one-parameter subgroup generated by $A$.*

*Proof.* There are three cases:

CASE 1. Suppose $A$ has a fixed point in $\Delta$. We may assume the fixed point is 0, so that $A(z) = e^{i\theta}z$. Then $e^{i\theta}B(0) = (AB)(0) = (BA)(0) = B(0)$. Since $A$ is not the identity, $B(0) = 0$, and $B$ has the form $e^{i\phi}z$.

CASE 2. Suppose $A$ has two fixed points on $\{|z| = 1\}$ which are different. We can map the problem to the right half-plane with the fixed points going to 0 and $\infty$, and $A(z) = \lambda z$ for some $\lambda > 0$, $\lambda \neq 1$. As above, $B$ either fixes each of $0, \infty$ or interchanges them. In the second case, $B(z) = \mu/z$ for some $\mu > 0$ and does not commute with $A$. Therefore $B$ fixes these points and $B(z) = \mu z$.

CASE 3. Suppose $A$ has one fixed point on $\{|z| = 1\}$. Again map the problem to the right half-plane with $\infty$ fixed. Then one sees easily that $A(z) = z + \lambda i$ for some real $\lambda \neq 0$ and $B(z) = z + i\mu$ for some $\mu \in \mathbf{R}$. $\square$

Möbius transformations corresponding to these three cases are called *elliptic, hyperbolic,* and *parabolic,* respectively. It is easy to show that every such transformation preserving the disk (except the identity) is one of these three types.

*Proof of Lemma 2.3 (continued).* Choose $h \in \Gamma_0$ which is not the identity. Since $\mathcal{G}$ commutes with $h$ it belongs to the one-parameter group generated by $h$. Since $\mathcal{G}$ is discrete and infinite we conclude that $\mathcal{G}$ has the form $\{g^n\}_{-\infty}^{\infty}$. This means that the fundamental group of $U$ is isomorphic to the integers, and $U$ is doubly connected. Since $U$ is hyperbolic, $U$ cannot be a punctured plane, and $U$ is either an annulus or a punctured disk. One of the alternatives (2), (4) or (5) must hold. $\square$

There is one more technical obstacle to surmount before completing the proof of the classification theorem. It is handled by the following variant of the "snail lemma". The reason for the nomenclature will become apparent from the proof, which stems from Section 54 of [**Fa2**].

LEMMA 2.4. *Suppose $U$ is hyperbolic, $f : U \rightarrow U$ is analytic on $U$ and across $\partial U$, and $f^n(z_0) \rightarrow \partial U$ for some $z_0 \in U$. Then there is a fixed point $\zeta \in \partial U$ for $f$ such that $f^n(z) \rightarrow \zeta$ for all $z \in U$. Either $\zeta$ is an attracting fixed point, or $\zeta$ is a parabolic fixed point with multiplier $f'(\zeta) = 1$.*

*Proof.* By Theorem I.4.3, the spherical distances between the iterates $z_n$ and $z_{n+1}$ of $z_0$ tend to 0. Thus the limit set of $\{z_n\}$ is a connected subset of $\partial U$, and furthermore, by continuity of $f$, any limit point is a fixed point for $f$. Since the fixed points of $f$ are isolated, we conclude that $z_n \rightarrow \zeta$ for some fixed point $\zeta \in \partial U$ of $R$. The orbit of every other $z \in U$ also converges to $\zeta$, since it remains a bounded hyperbolic distance from the orbit of $z_0$. The fixed point $\zeta$ is not repelling, since $z_n \rightarrow \zeta$.

Suppose $f'(\zeta) = e^{2\pi i\theta}$ where $\theta$ is rational. Then $U$ is contained in the basin of attraction associated with one of the petals at $\zeta$. Since the local rotation $f$ at $\zeta$ induces a cyclic permutation of the petals at $\zeta$, and since $f$ leaves $U$ invariant, in fact $f$ induces the identity permutation, and $f'(\zeta) = 1$.

Suppose $f'(\zeta) = e^{2\pi i\theta}$, where $\theta$ is irrational. Assume $\zeta = 0$. Let $z_0 \in U$, and let $V$ be a relatively compact subdomain of $U$ such that $V$ is simply connected and $z_0$ and $z_1 = f(z_0)$ belong to $V$. Since $f$ is univalent near 0 and $f^n \to 0$ uniformly on $V$, we can assume each $f^n$ is univalent on $V$. Then

$$\varphi_n(z) = \frac{f^n(z)}{f^n(z_0)}, \qquad z \in V,$$

is also univalent on $V$, $\varphi_n(z_0) = 1$, and $0 \notin \varphi_n(V)$. Let $\psi$ be the Riemann map from $\Delta$ to $V$, $\psi(0) = z_0$. Then $h_n(\zeta) = \varphi_n(\psi(\zeta)) - 1$ is univalent on $\Delta$, $h_n(0) = 0$, $h_n'(0) = \varphi_n'(z_0)\psi'(0)$, and $h_n$ omits $-1$. Thus the function $h_n/h_n'(0)$ belongs to $S$ and omits $-1/h_n'(0)$. The Koebe one-quarter theorem implies $|h_n'(0)| \leq 4$. Since $S$ is a normal family, the sequence $\{h_n/h_n'(0)\}$ is normal on $\Delta$, as is $\{h_n\}$. Consequently $\{\varphi_n\}$ is normal on $V$.

We claim that all limit functions of $\{\varphi_n\}$ are nonconstant. For if $|\varphi_n - 1| < \delta$ then $f^n(V)$ would be included in a narrow angle with vertex 0, and if this angle were smaller than $\theta/3$ then since $f(z) \cong e^{i\theta}z$ near 0, $f^{n+1}(V)$ would be disjoint from $f^n(V)$, contrary to hypothesis. Thus $\varphi_n'(z_0)$ is bounded away from 0. From Koebe's theorem again we deduce that there is $\rho > 0$ such that $\varphi_n(V)$ contains a disk centered at 1 of radius $\rho$. Thus $f^n(V)$ contains a disk centered at $z_n$ of radius $\rho|z_n|$.

Choose $N$ so that the disks of radius $\rho/2$ centered at the points $e^{2\pi i m\theta}$, $0 \leq m \leq N$, cover an annulus containing the unit circle. Since $z_{m+1} = e^{2\pi i\theta}z_m + o(|z_m|)$, the disks centered at $z_m$ of radius $\rho|z_m|$, $n \leq m \leq n+N$, cover an open annulus containing $z_n$ and $z_{n+1}$ for $n$ large. Hence $\cup f^n(V)$ contains a punctured neighborhood of 0, and 0 is an isolated point of $\partial U$. But then Theorem II.6.2 implies that $f$ is conjugate to a rotation about 0, contradicting $f^n(z) \to 0$. We conclude that $f'(\zeta)$ cannot be irrational. $\square$

*Proof of Theorem 2.1 (continued).* The proof is now complete, in view of the following observations. Since $R^n$ has degree $> 1$, no power of $R$ can coincide with the identity, and case (2) of Lemma 2.3 is ruled out. Since $\mathcal{J}$ has no isolated points, case (5) of Lemma

2.3 is also impossible for $R$. Finally, since there are no attracting periodic points in $\mathcal{J}$, Lemma 2.4 produces a parabolic fixed point in $\partial U$ whenever $f^n(z) \to \partial U$. $\square$

EXAMPLE. Suppose $R$ is a finite Blaschke product of degree $d \geq 2$. We have seen in Section III.1 that the Julia set $\mathcal{J}$ of $R$ is either the entire unit circle or a totally disconnected subset of the circle. There are now four possibilities. If $R$ has an attracting fixed point on $\partial\Delta$, or if $R$ has a parabolic fixed point on $\partial\Delta$ with only one attracting petal, then $\mathcal{J}$ is totally disconnected. If $R$ has an attracting fixed point $\zeta \in \Delta$, then $1/\bar{\zeta} \in \overline{\mathbf{C}}\backslash\overline{\Delta}$ is also an attracting fixed point, and $\mathcal{J}$ separates the basins of attraction, so $\mathcal{J} = \partial\Delta$. The only remaining possibility is that $R$ has a parabolic fixed point on $\partial\Delta$ with two attracting petals, in which case again $\mathcal{J} = \partial\Delta$. One can check that each of these four cases already occurs for Blaschke products of degree two. In all cases, the periodic points of $R$ that are not fixed are repelling.

## 3.   The Wolff–Denjoy Theorem

Before leaving this area let us prove the following beautiful related theorem of J. Wolff (1926) and A. Denjoy (1926). This brief proof was discovered by A.F. Beardon (1990).

THEOREM 3.1. *Let $f : \Delta \to \Delta$ be analytic, and assume $f$ is not an elliptic Möbius transformation nor the identity. Then there is $\alpha \in \overline{\Delta}$ such that $f^n(z) \to \alpha$ for all $z \in \Delta$.*

*Proof.* The theorem is easy if $f$ is a Möbius transformation, and so we assume that $f$ is not an isometry with respect to the hyperbolic metric, and further that $f(0) \neq 0$. If the orbit of 0 visits any compact set infinitely often, Lemma 2.2 provides a fixed point in $\Delta$. Thus we may assume that the orbit of 0 accumulates on $\partial\Delta$. The problem is to show that the orbit can accumulate at only one point $\alpha$ of $\partial\Delta$. This is again easy if $f$ extends continuously to $\partial\Delta$. The main point of the theorem is that no continuity is assumed.

Define $f_\varepsilon(z) = (1-\varepsilon)f(z)$, which maps to a compact subset of $\Delta$. Let $z_\varepsilon$ be the fixed point of $f_\varepsilon$, and let $D_\varepsilon$ be the hyperbolic disk centered at $z_\varepsilon$ with radius $\rho(0, z_\varepsilon)$. Since $f_\varepsilon$ is contracting, $f_\varepsilon(D_\varepsilon) \subset$

$D_\varepsilon$. Now $D_\varepsilon$ is a euclidean disk with 0 on its boundary. Any limit $D$ of the $D_\varepsilon$'s is a euclidean disk with 0 on its boundary, and $f(D) \subset D$. Thus the point of tangency of $D$ and $\partial\Delta$ is the only possible limit point on $\partial\Delta$ of the orbit of 0. $\square$

Another approach to the Wolff–Denjoy theorem provides an illuminating application of the Herglotz formula. We replace $\Delta$ by the upper half-plane $\mathbf{H}$. We are now considering $f = u+iv$ as a mapping of $\mathbf{H}$ to itself, with $v = \operatorname{Im} f > 0$, and we must show that either every orbit tends to $\infty$, or else every orbit is bounded. The Herglotz representation of the positive harmonic function $v$ in the upper half-plane is

$$v(x,y) = cy + y \int_{-\infty}^{\infty} \frac{d\sigma(s)}{(x-s)^2 + y^2}, \qquad y > 0, \qquad (3.1)$$

where $c \geq 0$ and $\sigma$ is a positive measure satisfying

$$\int_{-\infty}^{+\infty} \frac{d\sigma(s)}{1 + s^2} < \infty.$$

We may assume $\sigma \neq 0$.

Suppose $c \geq 1$. From (3.1) we have $v(x,y) > y$, so that the half-planes $\{y > y_0\}$ are invariant under $f$. These correspond to the invariant tangent disk $D$ to $\partial\Delta$ in the preceding proof. Let $z_n = x_n + iy_n$ be an orbit. Then either $y_n$ increases to $+\infty$, or else $y_n$ increases to some finite limit value $y_\infty$, in which case (3.1) shows that $|x_n| \to \infty$. In any event $|z_n| \to \infty$.

Now suppose $0 \leq c < 1$. In this case orbits are bounded. To see this, we use the Herglotz formula for $f$,

$$f(z) = cz + \int_{-\infty}^{+\infty} \frac{d\sigma(s)}{s-z} + b, \qquad z \in \mathbf{H},$$

where $b$ is real. The problem is to obtain an estimate of the form $|f(z) - cz| = o(z)$ for large $|z|$. This can be done by expressing $f$ as

$$
\begin{aligned}
f(z) &= cz + \int_{|s| \geq A} \left[ \frac{1}{s-z} - \frac{1}{s} \right] d\sigma(s) + \int_{-A}^{A} \frac{d\sigma(s)}{s-z} + b' \\
&= cz + z \int_{|s| \geq A} \frac{d\sigma(s)}{s(s-z)} + \mathcal{O}(1)
\end{aligned}
$$

as $|z| \to \infty$ and by estimating carefully the integral that appears here.

# V

# Critical Points and Expanding Maps

Critical points and their forward orbits play a key role in complex dynamical systems. The forward orbit of the critical points is dense in the boundary of any Siegel disk and Herman ring. If the critical points and their iterates stay away from the Julia set, the mapping is expanding on the Julia set, and the Julia set becomes more tractable.

## 1. Siegel Disks

Let $CP$ denote the (finite) set of critical points of $R$. The *postcritical set* of $R$ is defined to be the forward orbit $\cup_{n\geq 0} R^n(CP)$ of the critical points. We denote by $CL$ the closure of the postcritical set of $R$. This set is important because on its complement all branches of $R^{-n}$, $n \geq 1$, are locally defined and analytic.

As we have already seen for polynomials (Section III.4), many basic properties of the dynamics and structure of $\mathcal{J}$ are determined by the critical points and the postcritical set. By Theorems III.2.2 and III.2.3, each attracting and parabolic cycle of components of the Fatou set $\mathcal{F}$ contains a critical point. Siegel disks and Herman rings contain no critical point, however in a certain sense they can be associated to critical points. We begin with the following theorem.

The proof is essentially that of Section 31 of [**Fa2**] and predates the establishment of the existence of rotation domains.

THEOREM 1.1. *If $U$ is a Siegel disk or Herman ring, then the boundary of $U$ is contained in the closure of the postcritical set of $R$.*

*Proof.* Let $U$ be a rotation domain that is invariant under $R$, and suppose that $CL$ does not contain $\partial U$. Let $D$ be an open disk disjoint from $CL$ which meets $\partial U$. We assume also that $D$ is disjoint from some open invariant subset $V \neq \emptyset$ of $U$. Define $f_n$ to be any branch of $R^{-n}$ on $D$. Since the $f_n$'s omit $V$, they form a normal family on $D$. Now $R$ is one-to-one on $U$, so there are other components of $R^{-1}(U)$. Since inverse iterates of any fixed point of $\mathcal{J}$ are dense in $\mathcal{J}$, there is for suitable $m \geq 1$ a component $W$ of $R^{-m}(U)$ distinct from $U$ that meets $D$. If $z \in D \cap W$, then $f_j(z)$ and $f_k(z)$ belong to different components of $\mathcal{F}$ for $j \neq k$, or else they would belong to a periodic component, which could not be iterated eventually to $W$ then $U$. Hence $f_k(z)$ tends to $\mathcal{J}$ for $z \in D \cap W$, and since $\mathcal{J}$ has no interior, any normal limit of the $f_k$'s is constant on $D \cap W$. On the other hand, since the $f_k$'s are rotations of $U$, any normal limit is nonconstant on $D \cap U$. This contradiction establishes the theorem. $\square$

THEOREM 1.2. *If the postcritical set of $R$ is finite and there are no superattracting cycles, then $\mathcal{J} = \overline{\mathbf{C}}$.*

*Proof.* Siegel disks and Herman rings require that $CL$ be infinite, since $CL$ is dense on the boundary. So does a parabolic component, since it contains a critical point and the iterates of this point are distinct. According to the remark after Theorem II.2.4, an attracting component that is not superattracting has a critical point with infinite forward orbit. Thus if $CL$ is finite, the only possibility for a periodic component of $\mathcal{F}$ is a superattracting cycle. By hypothesis, there are no such components, and so $\mathcal{F}$ is empty and $\mathcal{J} = \overline{\mathbf{C}}$. $\square$

EXAMPLE. Take $R(z) = 1 - 2/z^2$. The critical points of $R$ are $0$ and $\infty$, which are iterated as follows:

$$0 \rightarrow \infty \rightarrow 1 \rightarrow -1 \rightarrow -1 \rightarrow \cdots.$$

Thus $R$ can have no superattracting cycles, and $\mathcal{J} = \overline{\mathbf{C}}$.

THEOREM 1.3. *If there are two completely invariant components of $\mathcal{F}$, then these are the only components of $\mathcal{F}$.*

*Proof.* As remarked earlier (in the proof of Theorem IV.1.1), each completely invariant component of $\mathcal{F}$ contains $d - 1$ critical points, and there are no other critical points. Since rotation domains have no critical points, each completely invariant component is attracting or parabolic. Moreover, the iterates of each critical point tend to an attracting or parabolic cycle. In view of Theorem 1.1, there can be no Siegel disks or Herman rings, nor can there be other parabolic or attracting components, since they would require further critical points. Thus the completely invariant components are the only components of $\mathcal{F}$. □

We show in Section VI.4 that if there are two completely invariant components of $\mathcal{F}$, one of which has an attracting fixed point, then the Julia set $\mathcal{J}$ is a simple closed Jordan curve. It follows from Theorem VI.2.1 that if the two completely invariant components both have attracting fixed points, then $\mathcal{J}$ is a quasicircle.

EXAMPLE. For $0 < b < 1$, the function $R(z) = z/(z^2 - bz + 1)$ has a parabolic fixed point at $z = 0$ with multiplier $+1$ and an attracting fixed point at $b$ with multiplier $1 - b^2$. The attracting petal at 0 and the immediate basin of attraction at $b$ contain the critical points $-1$ and $+1$, respectively. Since $R$ has degree two and is two-to-one on each immediate basin of attraction, each immediate basin is completely invariant. By Theorem 1.3, there are no other components of $\mathcal{F}$. As mentioned above, $\mathcal{J}$ is a simple closed Jordan curve, passing through 0 and $\infty$. For $b = 0$, we have only the parabolic fixed point at 0, which has two completely invariant attracting petals. The respective petals include the right and left half-planes. Since the Julia set separates the petals, it coincides with the (extended) imaginary axis.

The classification theorem shows that the iterates of any point of $\mathcal{F}$ either remain in a compact subset of $\mathcal{F}$ or converge to a parabolic cycle in $\mathcal{J}$. It follows that the boundary $\partial U$ of any Siegel disk $U$ lies in the closure of the forward orbits of those critical points that lie on $\mathcal{J}$. We claim that in fact there is a single critical point in $\mathcal{J}$ whose forward orbit is dense in $\partial U$. Indeed, since there are only

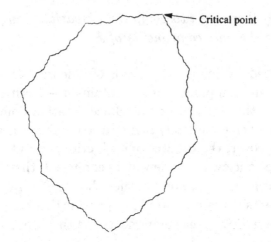

Critical point

FIGURE 1. Iterates of a critical point delineate a Siegel disk.

finitely many critical points, there is a critical point $q \in \mathcal{J}$ whose forward orbit is dense in a relatively open subset $W \neq \emptyset$ of $\partial U$. Replacing $W$ by $\cup R^n(W)$, we can assume that $W$ is invariant under $R$. Let $\psi : \Delta \to U$ be a conjugation of multiplication by $\lambda$ on $\Delta$ and $R$ on $U$. Let $S$ be the set of $\zeta \in \partial \Delta$ at which $\psi$ has a radial limit belonging to $W$. Since $W$ is open and nonempty, $S$ has positive length on $\partial \Delta$, and further $S$ is invariant under the rotation $\zeta \to \lambda \zeta$. If $\zeta_0$ is any point of Lebesgue density of $S$, then so is each rotate $\lambda^n \zeta_0$, and evidently every point of $\partial \Delta$ is a point of Lebesgue density of $S$, so that $S$ has full measure. It follows that $W$ is dense in $\partial U$, and the claim is established.

EXAMPLE. Consider $P(z) = \lambda(z - z^2/2)$ where $\lambda = \exp(2\pi i(\sqrt{5} - 1)/2)$. The golden mean $(\sqrt{5} - 1)/2$ is a favorite choice of rotation angle for obtaining computer-generated pictures of Siegel disks. From its continued fraction expansion

$$\frac{\sqrt{5} - 1}{2} = \cfrac{1}{1 + \cfrac{1}{1 + \cdots}}$$

it is the worst approximable number in the sense of Diophantine approximation (see p. 164 of [**HaW**]). By Siegel's Theorem II.6.4, the fixed point 0 of $P$ is the center of a Siegel disk. Figure 1 shows 20,000 iterates of the critical point 1 under $P$, which delineate the Siegel disk. (A similar picture of a Herman ring is given later in Section

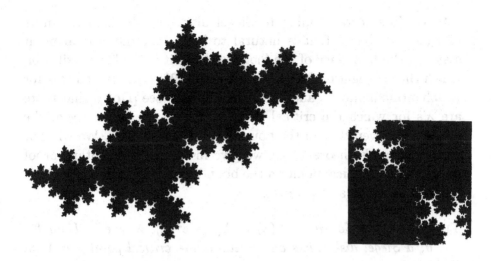

FIGURE 2. Siegel disk and enlargement at critical point.

VI.3.) On the left of Figure 2 is the filled-in Julia set for $P$, and on the right an enlargement around the critical point 1. An interesting question concerns whether the critical point is an accessible boundary point of $A(\infty)$ or not, that is, the end of a curve in $A(\infty)$. This seems to be the case in Figure 2.

Let $R(z) = \lambda z + a_2 z^2 + \cdots$, where $0 < |\lambda| < 1$, and suppose $z = h(\zeta)$ conjugates the multiplication $\zeta \to \lambda\zeta$ to $R(z)$. If $\varphi$ is Koenigs' coordinate function (appropriately normalized), then $h$ is the branch of $\varphi^{-1}$ mapping 0 to 0. Schröder's equation for the conjugation assumes the form

$$h(\zeta/\lambda) = R^{-1}(h(\zeta)), \qquad |\zeta| < \varepsilon. \qquad (1.1)$$

Since $|\lambda| < 1$, we can use this formula to continue $h$ until we meet a singularity of $R^{-1}$. Thus there is a critical point of $R$ on the boundary of the domain corresponding to the largest disk on which the conjugation holds (see the discussion in Section III.4). Using (1.1), it is easy to see that the boundary of this domain is an analytic curve except at the critical point. Here there is a corner of angle $\pi/m$, where $m-1$ is the order of the critical point. For $P_\lambda(z) = \lambda(z - z^2/2)$ with $|\lambda| < 1$, the angle is 90°. On the other hand, for $\lambda = \exp(2\pi i(\sqrt{5} - 1)/2)$ the angle seems to be very close to 120° (see Figure 2, and also the approximation on p. 561 of [MaN]), so there is some sort of discontinuity here. No rigorous results exist on this phenomenon.

Since $P_\lambda$'s corresponding to Siegel disks are obtained as limits of $P_\lambda$'s with $|\lambda| < 1$, it is natural to hope that the critical point stays on the boundary of the Siegel disk, as it is the largest disk on which the conjugation holds. M. Herman (1985) has proved this for Diophantine numbers, and he has also shown (see [Do3]) that there are $\lambda$'s for which the critical point is not on the boundary of the Siegel disk (even though the boundary is a quasicircle). We will not prove Herman's theorem, but we give instead an elementary proof that there is a critical point on the boundary for almost all quadratic polynomials with a Siegel disk.

THEOREM 1.4. *Consider* $P_\lambda(z) = \lambda(z - z^2/2)$, $\lambda = e^{i\theta}$. *Then for a.e.* $\theta$, *a Siegel disk exists and contains the critical point* $z = 1$ *on its boundary.*

*Proof.* We already know (Theorem II.6.4) that the conjugation exists for a.e. $\theta$, but we give a simpler proof for this case, due to Yoccoz. The proof of the theorem itself is a continuation of Yoccoz' proof (due to L. Carleson and P. Jones; see [Do3]).

For $0 < |\lambda| < 1$, consider the conjugating map $h(\zeta, \lambda)$ for $P_\lambda$, normalized so that $h(\zeta, \lambda)$ is univalent for $|\zeta| < 1$ and $h(1, \lambda)$ is the critical point 1 of $P_\lambda$. Write

$$h(\zeta, \lambda) = a_1(\lambda)\zeta + a_2(\lambda)\zeta^2 + \cdots, \qquad |\zeta| < 1, \ 0 < |\lambda| < 1.$$

Since $h$ omits 1, the Koebe one-quarter theorem gives $|h'(0, \lambda)| = |a_1(\lambda)| \leq 4$. This implies that $h(\zeta, \lambda)$ is an equicontinuous family of univalent mappings, and if we adjoin the function 0, we have a compact family. We study the behavior of $h(\zeta, \lambda)$ as $|\lambda| \to 1$.

Let $\varphi_\lambda$ be Koenigs' coordinate function, normalized by $\varphi'_\lambda(0) = 1$. Then $\varphi_\lambda$ depends analytically on $\lambda$. Since conjugating functions are unique up to dilations, and since $h^{-1}(1, \lambda) = 1$, we have $h^{-1}(z, \lambda) = \varphi_\lambda(z)/\varphi_\lambda(1)$. Thus $h^{-1}$, and also $h$, depend analytically on $\lambda$. The coefficient estimate of Theorem I.1.8 gives $|a_k(\lambda)| \leq ek^2|a_1(\lambda)| \leq 4ek^2$, so each $a_k(\lambda)$ extends analytically to 0, and the series expansion extends $h$ analytically to the bidisk $|\zeta| < 1, |\lambda| < 1$. Furthermore, the coefficient estimates, or the distortion theorem, show that for each fixed $\zeta$, $h(\zeta, \lambda)$ is bounded in $\lambda$. (It is easy to show that $|h| \leq 4$.)

The analyticity of $h(\zeta, \lambda)$ in $\lambda$ can also be seen in a more elementary way, following Yoccoz, by considering the sequence of polynomials $Q_n(\lambda) = \lambda^{-n} P_\lambda^n(1)$. Thus $Q_1 = 1/2$, $Q_2 = 1/2 - \lambda/8$,

$Q_3 = 1/2 - \lambda/8 - \lambda^2/8 + \lambda^3/16 - \lambda^4/128, \ldots$, and in general $Q_{n+1} = Q_n - \lambda^{n-1}Q_n^2/2$. One shows directly that $|Q_n(\lambda)| \leq 4$ for $|\lambda| < 1$, and $Q_n(\lambda) \to a_1(\lambda)$.

Let $E$ be the set of $e^{i\theta} \in \partial\Delta$ such that $a_1(\lambda) \to 0$ as $\lambda \to e^{i\theta}$. Then $E$ has zero measure. (This is an elementary fact, which can be proved by applying Fatou's lemma to the integrals of $\log|a_1(re^{i\theta})|$ as $r \to 1$.) Suppose $e^{i\theta} \notin E$. Choose $\lambda_k \in \Delta$ such that $\lambda_k \to e^{i\theta}$ and $a_1(\lambda_k) \to a \neq 0$. Let $\psi(\zeta)$ be any limit of the univalent functions $h(\zeta, \lambda_k)$. Then $\psi'(0) = a$, and $\psi$ is univalent. The functional equation for the conjugation $h$ is

$$h(\lambda\zeta, \lambda) = P_\lambda(h(\zeta, \lambda)), \qquad |\zeta| < 1, \ |\lambda| < 1,$$

and passing to the limit we obtain

$$\psi(e^{i\theta}\zeta) = P_{e^{i\theta}}(\psi(\zeta)), \qquad |\zeta| < 1.$$

Thus there is a Siegel disk centered at 0 for $P_{e^{i\theta}}$, and we have Siegel disks for almost all $\theta$.

Now conjugations are uniquely determined by $\psi'(0) = a$, so that $h(\zeta, \lambda_k)$ actually converges normally just as soon as $\lambda_k \to e^{i\theta} \in \partial\Delta$ and $a_1(\lambda_k)$ converges. In particular, if $a_1(\lambda)$ has a radial limit $a_1(e^{i\theta}) \neq 0$, then the conjugations $h(\zeta, \lambda)$ have a radial limit, which we denote by $h(\zeta, e^{i\theta})$. It determines a Siegel disk for $P_{e^{i\theta}}$.

We now want to prove that the critical point 1 remains on the boundary of almost all the Siegel disks. We study the functions

$$u(r, \lambda) = \log \frac{1}{|h(r, \lambda) - 1|}, \qquad |\lambda| < 1, \ 0 < r < 1.$$

For $r$ fixed, $u$ is harmonic for $\lambda$ in $\Delta$ and bounded below. By our discussion of the conjugation for $|\lambda| < 1$ preceding the statement of the theorem, we know $h$ maps $\Delta$ onto a smooth domain with a $\pi/2$ corner at the critical point 1. This implies

$$u(r, \lambda) > \frac{1}{2}\log\frac{1}{1-r} + C_0$$

for all $r > r(\lambda)$. In order to place the critical point 1 on the boundary of the Siegel disk, we want to show that 1 is a limit point of $h(r, e^{i\theta})$ as $r \to 1$. For this, it suffices to show $\limsup u(r, e^{i\theta}) = \infty$. Thus the estimate above is what we need, except that it is not uniform in $\lambda$.

For $\lambda$ fixed, $0 < |\lambda| < 1$, the function $1 - h(z, \lambda)$ can be expressed as the quotient of two functions in $\mathcal{S}$, namely, $h(z, \lambda)/a_1(\lambda)$ and $h(z, \lambda)/[a_1(\lambda)(1 - h(z, \lambda))]$. The upper and lower estimates in the distortion theorem (Theorem I.1.6) then yield

$$|1 - h(z, \lambda)| \geq \frac{(1 - |z|)^2}{|z|} \cdot \frac{|z|}{(1 + |z|)^2} \geq \frac{(1 - |z|)^2}{4}.$$

Hence

$$u(r, \lambda) \leq 2 \log \frac{1}{1 - r} + \log 4.$$

This estimate persists for the radial boundary values at $\lambda = e^{i\theta}$.

Fix $0 < c < 1/2$, take a sequence $r_k \to 1$, and define

$$A_N = \{\theta : |u(r_k, e^{i\theta})| > c \log \frac{1}{1 - r_k} \text{ for some } k \geq N\}.$$

It suffices to show that $A_N$ has full measure for all $N$. Indeed, then $\cap A_N$ has full measure, and for almost all $\theta$ we obtain a sequence of $r$'s tending to 1 such that $h(r, e^{i\theta}) \to 1$.

For arbitrary $\theta_0$ consider $\lambda = \rho e^{i\theta_0}$. Fix $M$ large, and let $I$ be the interval of length $M(1 - |\lambda|)$ centered at $e^{i\theta_0}$. The Poisson kernel $P_\lambda(\theta) = (1 - |\lambda|^2)/|e^{i\theta} - \lambda|^2$ satisfies

$$\int_E P_\lambda(\theta) \frac{d\theta}{2\pi} \leq c_1(M) \frac{|E \cap I|}{|I|} + \frac{c_2}{M}$$

for any subset $E$ of $\partial\Delta$, where $|\cdot|$ denotes angular measure. Choose $k$ so large that $k \geq N$ and $r_k > r(\lambda)$. From the Poisson formula we obtain

$$
\begin{aligned}
\frac{1}{2} \log \frac{1}{1 - r_k} + C_0 \;\leq\; & u(r_k, \lambda) = \int P_\lambda(\theta) u(r_k, e^{i\theta}) \frac{d\theta}{2\pi} \\
\leq\; & c \log \frac{1}{1 - r_k} + \int_{A_N} P_\lambda(\theta) u(r_k, e^{i\theta}) \frac{d\theta}{2\pi} \\
\leq\; & c \log \frac{1}{1 - r_k} + 2 \log \frac{1}{1 - r_k} \int_{A_N} P_\lambda(\theta) \frac{d\theta}{2\pi} + \log 4 \\
\leq\; & \log \frac{1}{1 - r_k} \left[ c + 2 c_1(M) \frac{|A_N \cap I|}{|I|} + \frac{2 c_2}{M} \right] + \log 4.
\end{aligned}
$$

Solving for $|A_N \cap I|/|I|$ and letting $r_k \to 1$, we obtain

$$\frac{|A_N \cap I|}{|I|} \geq \frac{1}{2 c_1(M)} \left( \frac{1}{2} - c - \frac{2 c_2}{M} \right) \geq c_0 > 0$$

for some suitable $c_0$. Hence $A_N$ has positive density everywhere. By Lebesgue's differentiation theorem, $A_N$ has full measure. $\square$

## 2.   Hyperbolicity

A rational function $R$ is *hyperbolic* (on $\mathcal{J}$) if there is a metric $\sigma(z)|dz|$ smoothly equivalent to the spherical metric in a neighborhood of $\mathcal{J}$, with respect to which $R$ is expanding. If $\infty \notin \mathcal{J}$, this means that

$$\sigma(R(z))|R'(z)| \geq A\sigma(z), \qquad z \in \mathcal{J}, \tag{2.1}$$

for some fixed $A > 1$. Then the identity

$$\frac{\sigma(R^n(z))(R^n)'(z)}{\sigma(z)} = \prod_{k=0}^{n-1} \frac{\sigma(R^{k+1}(z))}{\sigma(R^k(z))} R'(R^k(z))$$

shows that $\sigma(R^n(z))|(R^n)'(z)| \geq A^n\sigma(z)$ for all $n \geq 1$.

LEMMA 2.1. *Suppose* $\infty \notin \mathcal{J}$. *Then the following are equivalent.*

1. $R$ *is hyperbolic on* $\mathcal{J}$.

2. *There exist* $a > 0$ *and* $A > 1$ *such that* $|(R^n)'| \geq aA^n$ *on* $\mathcal{J}$ *for all* $n \geq 1$.

3. *There exists* $m \geq 1$ *such that* $|(R^m)'| > 1$ *on* $\mathcal{J}$, *that is, such that* $R^m$ *is expanding on* $\mathcal{J}$ *with respect to the euclidean metric.*

*Proof.* To show (1) implies (2), set $a = \min \sigma / \max \sigma$ and apply the estimate preceding the lemma. Clearly (2) implies (3). If (3) holds, then

$$\sigma(z) = |R'(R^{m-2}(z))|^{1/m} |R'(R^{m-3}(z))|^{2/m} \cdots |R'(z)|^{1-1/m}$$

defines a metric for which $R$ is expanding. $\square$

From condition (3) of the lemma, it is clear that if $R^n$ is hyperbolic on $\mathcal{J}$ for some $n$, then so is $R$, as are all iterates of $R$.

EXAMPLE. $R(z) = z^{\pm d}$ is expanding on $\mathcal{J} = \{|z| = 1\}$, since $|R'(z)| = d$ on $\mathcal{J}$.

Note that if $R$ is hyperbolic on $\mathcal{J}$, there can be no critical points on $\mathcal{J}$. This rules out the case $\mathcal{J} = \overline{\mathbf{C}}$, and it also rules out Siegel disks and Herman rings. Furthermore, $R$ can have no parabolic cycles, as these have multipliers of unit modulus. Thus if $R$ is hyperbolic on $J$, all components of $\mathcal{F}$ are iterated to attracting cycles. The converse is also true.

**THEOREM 2.2.** *The rational function $R$ is hyperbolic on $J$ if and only if the closure of the postcritical set of $R$ is disjoint from $J$. This occurs if and only if every critical point belongs to $\mathcal{F}$ and is attracted to an attracting cycle.*

*Proof.* Since the total multiplicity of the critical points is $2d - 2$, there are at least two points in $CL$. If there are exactly two, they are critical points of multiplicity $d - 1$. By placing these at $0$ and $\infty$, we see that $R$ is conjugate to $z^d$ or $z^{-d}$, hence hyperbolic.

We assume then that there are at least three points in $CL$ and that $CL \cap J = \emptyset$, so that each component of $\mathcal{F}$ iterates to an attracting cycle. Then $D = \overline{\mathbf{C}} \backslash CL$ is a hyperbolic domain, and all branches of $R^{-1}$ are analytically continuable along any path in $D$. Let $f$ be a lift to the universal covering space $D^\infty \cong \Delta$ of a locally defined branch of $R^{-1}$. Since the inverse iterates $R^{-k}(CL)$ are dense in $J$, $R^{-1}(CL)$ includes points of $D$, and these points are omitted by any branch of $R^{-1}$. Hence $f$ is a strict contraction with respect to the hyperbolic metric of $D^\infty$, and branches of $R^{-1}$ are locally strict contractions with respect to the hyperbolic metric of $D$. Thus $R$ is strictly expanding with respect to the hyperbolic metric on $D$, and $R$ is hyperbolic on $J$. $\square$

**THEOREM 2.3.** *If $R$ is hyperbolic, then $J$ has zero area.*

*Proof.* Assume $\infty \notin J$. Let $V$ be an open set containing $J$, invariant under $R^{-1}$, on which $1/R'$ and $R''$ are bounded, and on which $|(R^k)'| \geq a/c^k$ for some $0 < c < 1$. If $z$ and $w$ lie in a disk in $V$, and if $R^{-k}$ is any analytic branch on the disk, then

$$|R^{-k}(z) - R^{-k}(w)| \leq \frac{1}{a} c^k |z - w|.$$

Hence

$$\left| 1 - \frac{R'(R^{-k}(z))}{R'(R^{-k}(w))} \right| \leq C_0 |R^{-k}(z) - R^{-k}(w)| \leq C_1 c^k |z - w|.$$

Using the chain rule, we obtain

$$\frac{(R^{-n})'(w)}{(R^{-n})'(z)} = \prod_{k=0}^{n-1} \frac{R'(R^{-k}(z))}{R'(R^{-k}(w))} = \prod_{k=0}^{n-1}(1 + c^k|z - w|\mathcal{O}(1))$$
$$= 1 + |z - w|\mathcal{O}(1),$$

where the bounds are independent of $z, w, n$, and the branch of $R^{-n}$. This shows that branches of $R^{-n}$ are uniformly close to being affine on small disks.

Now fix $\varepsilon > 0$ small, and note that disks $\Delta(z, \varepsilon)$, with $z \in J$, have at least a fixed proportion (say $\delta > 0$) of their area in $\mathcal{F}$. Let $z_0 \in J$. Choose $n$ large, and set $z_n = R^n(z_0)$. Now $R^{-n}(\Delta(z_n, \varepsilon))$ is approximately a disk with at least $\delta/2$ of its area in $\mathcal{F}$, and its diameter tends to 0 as $n \to \infty$. Hence $z_0$ cannot be a point of full area density of $J$. By the Lebesgue differentiation theorem, $J$ has zero area. $\square$

It had been conjectured that for rational maps either $J$ has zero area or $J = \overline{\mathbf{C}}$. Recently S. van Strien and T. Nowicki showed that this conjecture already fails for a polynomial of the form $z^{16} + c$. There is a strong analogy with Kleinian groups, which is still not completely understood (see [Su2], [ErL]). The corresponding conjecture for Kleinian groups, that the limit set of a Kleinian group has zero area, is not completely settled.

## 3.   Subhyperbolicity

It is useful to consider a slightly weaker form of the notion of an expanding map. Assume $\infty \notin J$. Consider distances measured by $\sigma(z)|dz|$ where $\sigma \geq c > 0$, as above, and denote the corresponding distance function by $d_\sigma(z_1, z_2)$. We allow $\sigma$ to blow up at a finite number of exceptional points $a_1, ..., a_q$ but assume

$$\sigma(z) \leq \sum \frac{C}{|z - a_i|^\beta}$$

for some $\beta < 1$, so that $d_\sigma(z, a_i) < \infty$. Under these conditions, we say $\sigma$ is an *admissible metric*. Note that

$$c|z - w| \leq d_\sigma(z, w) \leq C|z - w|^{1-\beta}$$

for $z$ and $w$ near $\mathcal{J}$. We call $R$ *subhyperbolic* if $R$ is expanding on $\mathcal{J}$ for some admissible metric, that is, there is $A > 1$ such that (2.1) holds in a neighborhood of $\mathcal{J}$.

If $R$ is expanding on a neighborhood of $\mathcal{J}$ with respect to *any* metric (equivalent to the euclidean metric), then there are no neutral periodic points in $\mathcal{J}$. Indeed any periodic point $z_0$ then has an open neighborhood that is mapped to a compact subset of itself by the appropriate power $R^{-m}$, and this implies $|(R^m)'(z_0)| > 1$. In the subhyperbolic case, there are further restrictions.

THEOREM 3.1. *Suppose $\mathcal{J} \neq \overline{\mathbf{C}}$. Then $R$ is subhyperbolic if and only if each critical point in $\mathcal{J}$ has finite forward orbit, while each critical point in $\mathcal{F}$ is attracted to an attracting cycle.*

*Proof.* Suppose $R$ is subhyperbolic. The condition (2.1) shows that each critical value in $\mathcal{J}$ is an exceptional point for the metric $\sigma$, as is any iterate of an exceptional point. Hence $CL \cap \mathcal{J}$ is finite. By Theorem V.1.1, there are no rotation domains in $\mathcal{F}$. By the remark preceding the theorem, there are no parabolic periodic points. We conclude that $\mathcal{F}$ consists only of basins of attraction for attracting cycles.

For the converse, assume the condition on $CL$ holds. Then all the components of $\mathcal{F}$ are associated with attracting cycles, and there is an open neighborhood $V$ of $\mathcal{J}$ such that $R^{-1}(\overline{V}) \subset V$ and $V \cap CL \subset \mathcal{J}$. We are assuming $\infty \notin \mathcal{J}$. Let $\{a_1, ..., a_q\}$ be the critical values in $\mathcal{J}$ and their successive iterates. We build an $N$-sheeted branched covering surface $W$ over $V$ branched over the $a_j$'s so that any branch of $R^{-1}$, defined locally from $W$ to $W$, can be continued analytically along any path in $W$. For this there should be $N/n(a_j)$ branch points over $a_j$, each of order $n(a_j)$. The integers $n(a_j)$ are chosen so that, if $a \in R^{-1}(a_j)$ and $\nu(a)$ is the order of $R$ at $a$, then $n(a_j)$ is an integral multiple of $n(a)\nu(a)$. A moment's reflection shows this is always possible, the key point being that the cycles in $CL \cap \mathcal{J}$ contain no critical point or we would have a superattracting cycle. (If for instance there is only one critical point in $\mathcal{J}$, of order $n$, take $W$ to be $n$-sheeted, with a single branch point of order $n$ over each $a_j$.)

The universal covering surface of $W$ is the open unit disk $\Delta$, by Theorem I.3.3. Any locally defined inverse branch of $R^{-1}$ on $W$ lifts to an analytic map of $\Delta$ onto a proper subset. Hence branches of $R^{-1}$

are locally strict contractions with respect to the hyperbolic metric of $W$.

Parametrize the $k$th sheet of $W$ by $z$ and express the hyperbolic metric there as $\rho_k(z)|dz|$. Define $\sigma(z) = \sum \rho_k(z)$. Then $\sigma$ is smooth except at the branch points, and $\sigma(z) \sim |z - a_j|^{-1+1/n(a_j)}$ near $a_j$. The branches of $R^{-1}$ are strictly contracting with respect to each $\rho_k(z)|dz|$, hence with respect to $\sigma(z)|dz|$, so $R$ is strictly expanding. □

The inverse $R^{-1}$ of $R$ remains strictly contracting with respect to any small perturbation of the hyperbolic metric on $W$. If we perturb $\rho$ to be a constant times $|dw|$ at branch points, where $w$ is the local coordinate satisfying $z - a_j = w^{\nu(a_j)}$, and then sum over sheets, we obtain an admissible metric $\sigma$ for which $R$ is strictly expanding, such that $\sigma$ is smooth away from the $a_j$'s, and $\sigma(z)$ is a constant multiple of $|z - a_j|^{-1+1/\nu(a_j)}$ near $a_j$.

EXAMPLE. Consider $P(z) = z^2 + i$. Our metric should have singularities on the postcritical orbit $i \rightarrow i - 1 \rightarrow -i$. Define

$$\sigma(z) = \frac{1}{\sqrt{|z^2 + 1||z - (i - 1)|}},$$

so that $\sigma(P(z)) = |z|^{-1}|z^2 + 1|^{-1/2}|z^2 + 2i|^{-1/2}$. The inequality to be proved, $\sigma(P(z))|P'(z)| > A\sigma(z)$, can be simplified to $|z + (i - 1)| < 4/A^2$, which should hold for $z$ near $\mathcal{J}$. If $z \in \mathcal{J}$, then $|z| \leq \sqrt{3}$, and then $|z + (i - 1)| \leq \sqrt{3} + \sqrt{2} < 3.2$. So the inequality holds on $\mathcal{J}$ with $A = \sqrt{4/3.2} > 1.1$.

In Section VII.3 we extend the local affine estimate used in the proof of Theorem 2.3 to subhyperbolic maps. From this it follows that $\mathcal{J}$ has zero area also when $R$ is subhyperbolic.

## 4. Locally Connected Julia Sets

In this section we return to the case where $R = P$ is a polynomial of degree $d \geq 2$. Recall (Section III.4) that $A(\infty)$ is connected and $\mathcal{J} = \partial A(\infty)$. We are interested in determining when $\mathcal{J}$ is locally connected.

THEOREM 4.1. *If the polynomial $P$ is subhyperbolic on $\mathcal{J}$, and if $\mathcal{J}$ is connected, then $\mathcal{J}$ is locally connected.*

*Proof.* Since $\mathcal{J}$ is connected, $A(\infty)$ is simply connected. By Theorems 3.1 and III.4.1, $A(\infty)$ contains no (finite) critical points. The conjugation of $P$ to $\zeta^d$ at $\infty$ extends to an analytic map $\varphi : A(\infty) \to \{|\zeta| > 1\}$ satisfying $\varphi(P(z)) = \varphi(z)^d$. For fixed $R > 1$ we parametrize closed curves $\gamma_n$ by

$$\gamma_n(e^{i\theta}) = \varphi^{-1}(R^{1/d^n} e^{i\theta}), \qquad 0 \le \theta \le 2\pi, \ n \ge 1.$$

The $\gamma_n$'s tend to $\mathcal{J}$ as $n \to \infty$, and $P$ is a $d$-to-1 cover of $\gamma_{n+1}$ over $\gamma_n$. Let $\sigma$ be an admissible metric for which $P$ is strictly expanding near $\mathcal{J}$. If $0 < c < 1$ is the contraction constant for $P^{-1}$ near $\mathcal{J}$, then for $n$ large,

$$
\begin{aligned}
d_\sigma(\gamma_n(e^{i\theta}), \gamma_{n+1}(e^{i\theta})) &\le cd_\sigma(P(\gamma_n(e^{i\theta})), P(\gamma_{n+1}(e^{i\theta}))) \\
&= cd_\sigma(\gamma_{n-1}(e^{i\theta d}), \gamma_n(e^{i\theta d})).
\end{aligned}
$$

This shows that $\gamma_n(e^{i\theta})$ converges uniformly as $n \to \infty$. The limit function maps the circle continuously onto $\mathcal{J}$. Hence $\mathcal{J}$ is locally connected, and in fact the conformal map $\varphi^{-1}$ extends to a continuous map of $\{|\zeta| \ge 1\}$ onto $A(\infty) \cup \mathcal{J}$. □

THEOREM 4.2. *Suppose $P$ is a polynomial with finite postcritical set, such that no finite critical point is periodic. Then the Julia set $\mathcal{J}$ is a dendrite, that is, a compact, pathwise connected, locally connected, nowhere dense set that does not separate the plane.*

*Proof.* No components for $\mathcal{F}$ except $A(\infty)$ are possible, and Theorem 4.1 applies. □

EXAMPLE. If $P(z) = z^2 - 2$, then $0 \to -2 \to 2 \to 2 \to \cdots$, so the Julia set is a dendrite. In fact, we already know $\mathcal{J} = [-2, 2]$. A more interesting case is $P(z) = z^2 + i$, where $0$ is iterated to a cycle of length two: $0 \to i \to i - 1 \to -i \to i - 1 \to \cdots$. See Figure 3.

A crucial ingredient of the proof of local connectedness is the existence of an expanding metric. When there are parabolic cycles, such a metric cannot exist in a full neighborhood of the Julia set. However, under certain conditions such metrics exist in the part of $A(\infty)$ bordering $\mathcal{J}$. We follow Exposé No. X of [**DH2**].

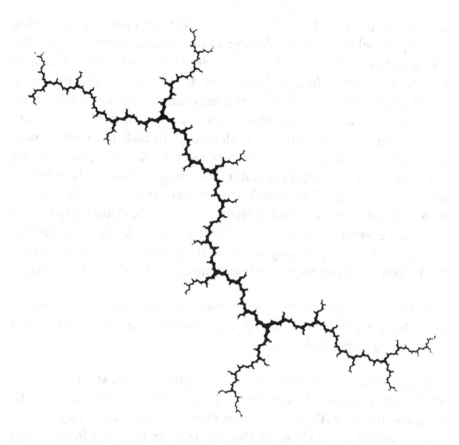

FIGURE 3. Julia set for $z^2 + i$.

For simplicity, suppose there is a parabolic fixed point $z_0 \in \mathcal{J}$ and that the forward iterates of each critical point either tend to an attracting cycle or to $z_0$ through an attracting petal. In view of the description in Section II.5 of the behavior of $P$ near a parabolic fixed point, we see that we can construct an open set $U$ containing $\mathcal{J} \backslash \{z_0\}$ such that $U$ is invariant under branches of $P^{-1}$, $U \cap A(\infty) = \Delta(0, R) \cap A(\infty)$ for some large $R$, and $U \cap \Delta(z_0, \varepsilon)$ is a finite union of narrow cusps terminating at $z_0$ and tangent to repelling directions. For $\varepsilon > 0$ sufficiently small, $P$ is expanding with respect to the euclidean metric $|dz|$ in each of these cusps, as can be seen from a simple calculation. Curves in $A(\infty)$ terminating at $z_0$ do not have finite length with respect to $d\rho_U$. To remedy this, we aim to modify $d\rho_U$ to be a multiple of $|dz|$ near $z_0$. This is done as follows.

Choose $M$ so large and $\varepsilon$ so small that $P(z)$ is expanding in a neighborhood of any point of $P^{-1}(\Delta(z_0, \varepsilon)) \cap U$ from the metric

$d\rho_U$ to the metric $M|dz|$. Set $V = P^{-1}(U)$, an open set containing $\mathcal{J}\setminus\{z_0\}$ on which $P$ is locally expanding with respect to $d\rho_U$. Define a metric $d\sigma(z)$ on $V$ by setting $d\sigma(z) = d\rho_U(z)$ on $V\setminus\Delta(z_0,\varepsilon)$ and $d\sigma(z) = \min\{d\rho_U(z), M|dz|\}$ on $V \cap \Delta(z_0,\varepsilon)$. The condition on $M$ guarantees that $P$ is locally expanding with respect to $\sigma$ on $V\setminus\Delta(z_0,\varepsilon)$. Since $P$ is expanding with respect to both $M|dz|$ and $d\rho_U$ in $V\cap\Delta(z_0,\varepsilon)$, it is expanding with respect to their minimum, hence it is locally expanding on $V$ with respect to $d\sigma$. Integrating along paths in $V$ and taking an infimum, we obtain a uniformly bounded metric $d_\sigma(z,w)$ on $V$ for which $P$ is strictly expanding (though not uniformly expanding). Such a metric can also be constructed when there are several parabolic cycles, and even when there are critical points on $\mathcal{J}$ so long as they are strictly preperiodic, as in Section 3. With such an expanding metric in hand, we obtain the following.

THEOREM 4.3. *Let $P$ be a polynomial with connected Julia set $\mathcal{J}$, such that each critical point of $P$ belonging to $\mathcal{J}$ is preperiodic. Then $\mathcal{J}$ is locally connected.*

*Proof.* Assume for convenience there is only one parabolic point $z_0$, with $p$ invariant petals. Let $d\sigma(z)$ be as above. Let the curves $\gamma_n$ be as in the proof of Theorem 4.1, so that $P$ is a $\sigma$-expanding map of each $\gamma_n$ onto $\gamma_{n-1}$. Let $\alpha_n$ be the geodesic arc in $A(\infty)$ from $\gamma_n(e^{i\theta})$ to $\gamma_{n+1}(e^{i\theta})$, and for $1 \le j < n$, set $\alpha_{n-j} = P^j(\alpha_n)$. Let $\varepsilon > 0$ be small. Suppose first that more than $3n/4$ of the $\alpha_j$'s meet $\Delta(z_0,\varepsilon)$. We arrange these in $k$ blocks of consecutive iterates, with $m_j$ arcs in the $j$th block. Then $k \le n/4$ and $m_1 + \cdots + m_k \ge 3n/4$. Estimating with the aid of the Fatou coordinate for a repelling direction at $z_0$, we see that for any fixed $0 < \delta < 1/p$, the $\sigma$-length of $\alpha_i$ is magnified by a factor at least $\mathcal{O}(m_j^{1+\delta})$ over the $j$th block. Thus the $\sigma$-length of $\alpha_n$ is at most $\mathcal{O}((m_1 \cdots m_k)^{-(1+\delta)})$ times the $\sigma$-length of $\alpha_1$. Now $m_1 \cdots m_k$ is minimized, subject to the constraints $m_j \ge 1$ and $m_1 + \cdots + m_k$ fixed, when all the $m_j$'s except one are 1, with minimum $m_1 + \cdots + m_k - (k-1) \ge n/2$. Thus the $\sigma$-length of $\alpha_n$ is at most $\mathcal{O}(n^{-(1+\delta)})$ times the length of $\alpha_1$. Suppose, on the other hand, that at least $n/4$ of the $\alpha_j$'s lie outside $\Delta(z_0,\varepsilon)$. On these $\alpha_j$'s, $P$ is uniformly expanding, so the $\sigma$-length of $\alpha_n$ is at most $C^{-n/4}$ times the length of $\alpha_1$, where $C > 1$ is an expansion constant. In any event we obtain $d_\sigma(\gamma_{n+1}(e^{i\theta}), \gamma_n(e^{i\theta})) = \mathcal{O}(n^{-(1+\delta)})$, the estimate being uniform in $\theta$. It follows that the curves $\gamma_n$ converge uniformly in the $\sigma$-metric. Since $d_\sigma(z_n, w_n) \to 0$ implies $|z_n - w_n| \to 0$, the $\gamma_n$'s converge also uniformly in the euclidean metric, and the limit $\mathcal{J}$ is locally connected. $\square$

EXAMPLE. Suppose the Fatou set of the polynomial $P$ consists of two completely invariant components $A(\infty)$ and $W$. Then all $d - 1$ finite critical points of $P$ belong to $W$, and the preceding theorem applies. The Riemann map from $\Delta$ to $A(\infty)$ extends to map $\partial\Delta$ continuously onto $\mathcal{J}$. Since $\mathcal{J}$ is adherent to $W$, the Riemann map must be one-to-one on $\partial\Delta$, and $\mathcal{J}$ is a simple closed Jordan curve. This applies for instance to the polynomial $z^2 + 1/4$, which has a parabolic fixed point at $1/2$ with one attracting petal. The Julia set of $z^2 + 1/4$ is the cauliflower set in Section VIII.1, Figure 5.

Lest the impression be left that $\mathcal{J}$ is always locally connected, we prove the following theorem of Douady and Sullivan (cf. [Su1]).

THEOREM 4.4. *Suppose the quadratic polynomial $P$ has an irrationally neutral fixed point that does not correspond to a Siegel disk. Then the Julia set $\mathcal{J}$ is a connected but not locally connected set.*

*Proof.* We use the result, proved in the next chapter (Theorem VI.1.2), that a polynomial of degree $d$ has at most $d-1$ attracting or neutral cycles (not including $\infty$). In the case at hand, there are no other attracting or neutral cycles than the irrationally neutral fixed point, so that $\mathcal{F} = A(\infty)$. We use also the fact that $A(\infty)$ is simply connected. By Theorem III.4.1, this is equivalent to the boundedness of $CL$. This in turn follows from Theorem VIII.1.3, and it can also be established by a direct argument.

We assume $z = 0$ is the irrationally neutral fixed point. Let $\varphi$ be the conjugation of $P$ to $\zeta^2$ at $\infty$. Since $A(\infty)$ has no finite critical points, $\varphi$ extends to a conformal map of $A(\infty)$ onto $\{|\zeta| > 1\}$, whose inverse $\psi : \{|\zeta| > 1\} \to A(\infty)$ satisfies $P(\psi(\zeta)) = \psi(\zeta^2)$.

Suppose $\psi$ extends continuously to $\{|\zeta| \geq 1\}$. Then $\psi^{-1}(0)$ is a closed subset of the unit circle $\{|\zeta| = 1\}$ of zero length. Let $\eta > 0$ be small, and let $D_\eta$ denote the set of $z \in A(\infty)$ such that $z = \psi(\zeta)$ for some $\zeta \in \Delta$ satisfying $\text{dist}(\zeta, \psi^{-1}(0)) < \eta$. Evidently $D_\eta$ is open, and the continuity of $\psi$ shows that $\overline{D_\eta}$ shrinks to $\{0\}$ as $\eta$ decreases to $0$. Thus we can assume that the branch $f$ of $P^{-1}$ satisfying $f(0) = 0$ is defined and analytic on a disk containing $D_\eta$.

We claim that $D_\eta$ is invariant under $f$. Indeed, let $z_0 \in D_\eta$, and write $z_0 = \psi(\zeta_0)$. Choose $\xi_0$, $|\xi_0| = 1$, such that $\psi(\xi_0) = 0$ and $|\zeta_0 - \xi_0| < \eta$. Let $z_1 = f(z_0)$. Then $z_1 = \psi(\zeta_1)$, where $\zeta_1^2 = \zeta_0$, $\zeta_1 = \sqrt{\zeta_0}$. Set $\xi_1 = \sqrt{\xi_0}$, where we use the same branch of the square root. Since $\zeta_1$ and $\xi_1$ are approximately in the same direction, $|\zeta_1 + \xi_1| > 1$. Hence $|\zeta_1 - \xi_1| < |\zeta_1 - \xi_1||\zeta_1 + \xi_1| = |\zeta_1^2 - \xi_1^2| = |\zeta_0 - \xi_0| < \eta$. Also $P(\psi(\xi_1)) = \psi(\xi_1^2) = \psi(\xi_0) = 0$. Hence $\psi(\xi_1)$ is either $0$ or $-\lambda$, and since $\xi_1$ is near $\zeta_1$ and $\psi(\zeta_1) = z_1$ is near $0$, $\psi(\xi_1) = 0$. Thus $z_1 \in D_\eta$, and $D_\eta$ is invariant under $f$.

Now $D_\eta$ includes $A(\infty) \cap \{|z| < \delta\}$ for some $\delta > 0$, by the uniform continuity of $\psi$. Thus if $z \in A(\infty), |z| < \delta$, then all iterates $f^m(z)$ are in $D_\eta$. Since $\mathcal{J} = \partial A(\infty)$ has no interior points and $\mathcal{F} = A(\infty)$, $D_\eta$ is dense in $\{|z| < \delta\}$. Thus by continuity, $f^m(z) \in \overline{D_\eta}$ for all $z, |z| < \delta$, and all $m \geq 1$. By Theorem II.6.2, $f$ can then be conjugated to a rotation, hence so can $P$, contrary to hypothesis. This contradiction shows that $\psi$ does not extend continuously to $\{|\zeta| = 1\}$. By Carathéodory's theorem, $\mathcal{J} = \partial A(\infty)$ is not locally connected. $\square$

# VI

# Applications of Quasiconformal Mappings

One of the basic ideas behind the use of quasiconformal mappings is to consider two dynamical systems acting in different parts of the plane and to construct a new system that combines the dynamics of both. This procedure is called *quasiconformal surgery*.

## 1. Polynomial-like Mappings

Let $U_1$ and $U_2$ be bounded, open, simply connected domains with smooth boundaries, such that $\bar{U}_1 \subset U_2$. Let $f(z)$ be holomorphic on $\bar{U}_1$ and map $U_1$ onto $U_2$ with $d$-fold covering, so that $f$ maps $\partial U_1$ onto $\partial U_2$, that is, $f$ is proper. Following Douady and Hubbard (1985), we call the triple $(f; U_1, U_2)$ *polynomial-like*. This name is motivated by the following.

THEOREM 1.1. *If $(f; U_1, U_2)$ is polynomial-like of degree $d$, then there are a polynomial $P$ of degree $d$ and a quasiconformal mapping $\varphi$ with $\varphi(z) = z + o(1)$ near $\infty$, such that $f = \varphi \circ P \circ \varphi^{-1}$ on $U_1$.*

*Proof.* We construct a Beltrami coefficient (ellipse field) for $\varphi$ as follows. Take $\rho > 1$. Let $\Phi(z)$ map $\mathbf{C} \backslash U_2$ conformally to $\{|\zeta| > \rho^d\}$ with $\Phi(\infty) = \infty$. We wish to extend $\Phi$ to $\mathbf{C} \backslash U_1$ so that $\Phi$ maps

$U_2 \backslash U_1$ one-to-one smoothly onto $\{\rho < |\zeta| < \rho^d\}$, so that for $z \in \partial U_1$,

$$\Phi(z)^d = \Phi(f(z)).$$

To do this, note that $f$ describes $\partial U_2$ $d$ times as $z$ runs through $\partial U_1$, so $\Phi$ can be defined on $\partial U_1$ and then extended arbitrarily in a smooth way to $U_2 \backslash U_1$. We define

$$g(z) = \begin{cases} f(z), & z \in U_1, \\ \Phi^{-1} \circ \Phi(z)^d, & z \in \mathbf{C} \backslash U_1. \end{cases}$$

Define an ellipse field by making it circles in $\mathbf{C} \backslash U_2$ and defining it to be $g$-invariant on

$$\bigcup_{n=1}^{\infty} g^{-n}(\mathbf{C} \backslash U_2).$$

This can be done uniquely as the sets $g(U_2) \backslash U_2, U_2 \backslash U_1, g^{-1}(U_2 \backslash U_1)$, ..., are disjoint, and each is mapped $d$-to-1 onto the preceding one by $g$. Furthermore, the circles are distorted only on the first iteration (inside $U_2 \backslash U_1$), since $g = f$ is analytic in $U_1$. The eccentricities of the ellipses do not change after the first iteration and so are uniformly bounded. Hence $|\mu| \leq k < 1$. We define $\mu \equiv 0$ on any remaining part of $\mathbf{C}$.

Let $\varphi$ solve the Beltrami equation corresponding to $\mu$. Then $P = \varphi \circ g \circ \varphi^{-1}$ maps infinitesimal circles to circles and is a $d$-to-1 mapping, hence a polynomial of degree $d$. Finally, $g = \varphi^{-1} \circ P \circ \varphi = f$ in $U_1$, as desired. $\square$

We can use Theorem 1.1 to prove the statement used in the proof of Theorem V.4.4.

THEOREM 1.2 (Douady). *A polynomial $P$ of degree $d$ has at most $d - 1$ nonrepelling cycles in the finite plane.*

*Proof.* Let $M$ be the set of points occurring in nonrepelling cycles. By Theorem III.2.7, $M$ is finite. Choose a polynomial $Q$ such that $Q(z) = 0$ for all $z \in M$, and $\sum \text{Re}\,(Q'(z_j)/P'(z_j)) < 0$ for all neutral cycles $\{z_1, ..., z_m\}$. To see that this is possible, note that $P'(z_j) \neq 0$ and that the conditions on $Q$ are linear, so the conditions can be satisfied if the degree of $Q$ is large enough. Suppose $\varepsilon > 0$ is small, and define $f = P + \varepsilon Q$. Then since $Q = 0$ on $M$, the nonrepelling cycles of $P$ are also cycles of $f$, and they must be attracting, since for any neutral cycle for $P$,

$$
\begin{aligned}
\sum \log |f'(z_j)| &= \sum \log |P'(z_j)| + \sum \log \left| 1 + \varepsilon \frac{Q'(z_j)}{P'(z_j)} \right| \\
&= 0 + \varepsilon \sum \mathrm{Re} \left( \frac{Q'(z_j)}{P'(z_j)} \right) + \mathcal{O}(\varepsilon^2) < 0
\end{aligned}
$$

if $\varepsilon$ is small enough. Hence all nonrepelling cycles for $P$ are attracting for $f$.

Assume $P$ is monic, take $\rho$ large, and set $U_2 = \Delta(0, \rho^d)$, $U_1 = f^{-1}(U_2)$. If $\varepsilon$ is sufficiently small, then $f \sim P \sim z^d$ near $\{|z| = \rho\}$, so $f$ is a $d$-to-1 proper mapping from $U_1$ to $U_2$, and $(f; U_1, U_2)$ is polynomial-like. We can assume $M$ is included in $U_1$. By Theorem 1.1, $f$ is conjugate to a polynomial $S$ of degree $d$, and every attracting cycle for $f$ must be attracting for $S$. By Theorem III.2.2, a polynomial has at most $d - 1$ attracting cycles (not counting $\infty$). Thus $P$ had at most $d - 1$ nonrepelling cycles. $\square$

Since $\infty$ is an attracting fixed point of order $d - 1$, Douady's theorem above is the same as Shishikura's theorem mentioned in Section III.2 in the case of a polynomial.

EXAMPLE. Suppose $P$ is a quadratic polynomial that has a cycle corresponding to a Siegel disk, $U_1 \to U_2 \to \cdots \to U_m \to U_1 \to \cdots$. The components of $\mathcal{F}$ in the cycle are mapped one-to-one onto themselves by $P$. Each $U_j$ in the cycle is the image of $U_{j-1}$ and one other bounded component $V_j$ of $\mathcal{F}$. Each $V_j$ has two distinct preimage components, each of these has two further distinct preimage components, and so on, so that the inverse images of each $V_j$ form a tree. By Theorem 1.2, there are no attracting or neutral cycles other than the Siegel cycle (and $\infty$). From the classification theorem, we conclude that each bounded component of $\mathcal{F}$ is eventually iterated by $P$ to the Siegel cycle.

## 2.  Quasicircles

A Jordan curve $\Gamma$ is called a *quasicircle* if it is the image of a circle under a quasiconformal homeomorphism of the sphere. Quasicircles can be geometrically characterized by the "three-point property":

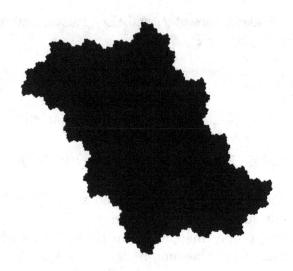

FIGURE 1. $P(z) = z^2 + i/2$, $\mathcal{J}$ is a quasicircle.

there is an $M > 0$ such that if $z_1, z_2, z_3 \in \Gamma$ and $z_2$ is on the "shorter" arc between $z_1$ and $z_3$, then $|z_1 - z_2| + |z_2 - z_3| \le M|z_1 - z_3|$. See [A2].

THEOREM 2.1 (Sullivan). *Suppose that the Fatou set of a rational function $R$ has two components and that $R$ is hyperbolic on the Julia set $\mathcal{J}$. Then $\mathcal{J}$ is a quasicircle.*

*Proof.* The proof is very similar to that of Theorem 1.1 (see also the proof of Theorem VII.5.1). Suppose the two components are $D_1$ and $D_2$. By considering $R^2$ if necessary, we may assume each component is completely invariant. They are simply connected, by Theorem IV.1.1. In $D_1$ construct a curve $\gamma_2$ with no critical points in the annular domain between $\gamma_2$ and $\mathcal{J}$. Let $\gamma_1 = R^{-1}(\gamma_2)$. Since $R$ is hyperbolic we can assume that $\gamma_1$ lies between $\gamma_2$ and $\mathcal{J}$. The domain between $\gamma_1$ and $\mathcal{J}$ corresponds to $U_1$, the domain between $\gamma_2$ and $\mathcal{J}$ to $U_2$. As before, we construct $g_1$ in $D_1$ and the analogous Beltrami coefficient on $D_1$. We do the same in $D_2$. We set $\mu = 0$ elsewhere and solve the Beltrami equation for $\varphi$. Again $R_0 = \varphi \circ g \circ \varphi^{-1}$ is analytic, and $\deg R_0 = \deg R = d$. Evidently $R_0$ has critical points of order $d - 1$ at two points, since $g_1$ and $g_2$ do, and hence is equivalent to $z^d$. Thus $\mathcal{J}$ is the image of the circle $\{|z| = 1\}$ under $\varphi$ composed with a Möbius transformation. $\square$

EXAMPLE. Suppose $P$ is a quadratic polynomial with an attracting fixed point. By Theorem III.2.2 the immediate basin of attraction $A^*$ of the fixed point contains the critical point, so $P$ maps $A^*$ two-to-one onto itself, and $A^*$ is completely invariant, as is $A(\infty)$. By Theorem V.1.3 these are the only components of $\mathcal{F}$. Furthermore, $P$ is hyperbolic on $\mathcal{J}$, by Theorem V.2.1. Thus Theorem 2.1 applies, and $\mathcal{J}$ is a quasicircle. An example is shown in Figure 1. This picture was produced by iterating $P$ and coloring black those points $z$ for which $|P^n(z)| \leq 4$ for $n = 1, 2, ..., 100$.

## 3.   Herman Rings

The first examples of Herman rings were produced by M. Herman (1984), who based their existence on Arnold's theorem (Theorem II.7.2). For $a > 1$ real, consider

$$R(z) = e^{2\pi i\theta}z^2(z-a)/(1-az).$$

It is an orientation-preserving homeomorphism of the unit circle, and it is uniformly close to the rotation $z \to e^{2\pi i\theta}z$ for large $a$. Now the rotation number of $R$ moves continuously with $a$ and $\theta$, it is increasing in $\theta$, and it converges uniformly to $\theta$ (mod 1) as $a \to +\infty$. For a fixed Diophantine number $\alpha$, we can therefore find $\theta = \theta(\alpha)$ so that $R$ has rotation number $\alpha$. Moreover $\theta(\alpha) \to \alpha$ as $a \to \infty$. Hence for large $a$ we have $R(z) = e^{2\pi i\alpha}z\Phi(z)$ where $\Phi$ is arbitrarily close to 1. The hypotheses of Theorem II.7.2 are met, and $R$ is conjugate to a rotation in some annulus containing the unit circle. The circle is contained in a fixed component $U$ of the Fatou set, and $U$ cannot be attracting or parabolic (since the circle is invariant) nor a Siegel disk (since 0 and $\infty$ are attracting). Thus $U$ must be a Herman ring. Actually, from the global version of Arnold's theorem we obtain a Herman ring for any $a > 1$ and any $\theta$ such that $R$ has a Diophantine rotation number. If for instance $a = 4$, the finite critical points of $R$ are the roots of $z(z^2 - (19/8)z + 1)$, which are $r_1 = 0$, $r_2 \approx 0.547$, and $r_3 \approx 1.828$. The first is superattracting, so by Theorem V.1.1, $\partial U$ must be contained in the closure of the iterates of the second and third. Figure 2 shows 20,000 iterates of each critical point for a randomly chosen value $\theta = 0.5431245$.

We give a second proof of the existence of Herman rings, due to

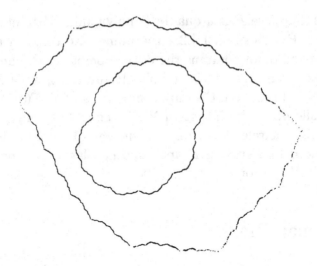

FIGURE 2. Iterates of critical points delineate a Herman ring.

M. Shishikura (1987), which transforms a Siegel disk into a Herman ring with the same rotation number.

THEOREM 3.1. *Suppose a rational function R has a Siegel disk, on which R is conjugate to multiplication by $\lambda = e^{2\pi i\theta}$. Then there exists another rational function with a Herman ring, on which it is conjugate to multiplication by the same multiplier $\lambda$.*

*Proof.* Let $U$ denote the Siegel disk of $R$, say with center 0, and let $\tilde{R}(z) = \overline{R(\bar{z})}$. Then $\tilde{R}$ has a Siegel disk $\tilde{U}$ with multiplier $e^{-2\pi i\theta}$ at 0. Thus $U$ and $\tilde{U}$ are conformally equivalent to the unit disk, and $R$ and $\tilde{R}$ act there by multiplication by $e^{2\pi i\theta}$ and $e^{-2\pi i\theta}$, respectively. For $0 < r, \tilde{r} < 1$ let $\gamma \subset U$ correspond to a circle $\{|\zeta| = r\}$ via the conjugating map, and let $\tilde{\gamma}$ be an analogous curve in $\tilde{U}$ corresponding to $\tilde{r}$. Define an orientation-reversing homeomorphism $\psi$ from an annular neighborhood of $\gamma$ to an annular neighborhood of $\tilde{\gamma}$ so that $\psi$ in $z$-coordinates corresponds to the map $\zeta \to r\tilde{r}/\zeta$ in $\zeta$-coordinates. Then $\psi(\gamma) = \tilde{\gamma}$, and $\psi(R(z)) = \tilde{R}(\psi(z))$ for $z \in \gamma$.

Extend $\psi$ to be a diffeomorphism from the domain inside $\gamma$ onto the domain outside $\tilde{\gamma}$. We may assume $\psi$ is analytic off an $R$-invariant annular neighborhood $V$ of $\gamma$, and that $\psi(V)$ is an $\tilde{R}$-invariant annular neighborhood of $\tilde{\gamma}$. Define a smooth function $g$ on $\overline{\mathbb{C}}$ by setting $g = R$ outside $\gamma$ and $g = \psi^{-1} \circ \tilde{R} \circ \psi$ inside $\gamma$. Then $g$ maps $V$ one-to-one onto $V$, and $g$ is analytic off $V$. Let $E$ be the ellipse field on

$V$ that is mapped to circles by $\psi$. As before, extend $E$ to $\bigcup g^{-n}(V)$ to be invariant under $g$, and extend $E$ to the remainder of $\overline{\mathbf{C}}$ to be circles.

Let $\varphi$ solve the Beltrami equation and observe that $R_0 = \varphi \circ g \circ \varphi^{-1}$ is rational, of degree $2d - 1$. The annular domain $W = \varphi(V)$ is mapped conformally onto itself by $R_0$, so $W$ is part of a component of $\mathcal{F}(R_0)$. This component is either a Siegel disk or a Herman ring. The first is impossible. Indeed $R_0^n$ cannot be normal in either complementary component of $W$, since $R^n$ and $\tilde{R}^n$ are not normal outside $U$ and $\tilde{U}$, respectively. Since the action of $R_0$ on $\varphi(\gamma)$ is conjugate to the action of $R$ on $\gamma$, and the rotation number of $R$ on $\gamma$ is $\theta$, the rotation number of $R_0$ on $\varphi(\gamma)$ is also $\theta$, and $R_0$ must be conjugate to multiplication by $e^{2\pi i\theta}$ on the Herman ring. $\square$

## 4.   Counting Herman Rings

By means of quasiconformal surgery, Shishikura showed there are at most $d - 2$ Herman rings, and this bound is sharp. Here we are concerned only with showing that the number of Herman rings is finite.

THEOREM 4.1. *The number of cycles of Herman rings of a rational function of degree $d$ is at most $4d + 2$.*

*Proof.* The proof is by counting parameters. The number of independent quasiconformal deformations of the system to nonconjugate systems cannot exceed the number of parameters describing the conjugacy classes. The line of argument is due to Sullivan (cf. [Do3]), and it is similar to that used in Theorem IV.1.3.

Let $U$ be a Herman ring, and let $E$ be a compact invariant subset of $U$ that is conformally equivalent to the annulus $\{1 \le |\zeta| \le R\}$. Consider the function $\nu$ on $E$ which in $\zeta$-coordinates is given by $\zeta^2/|\zeta|^2$. For $0 < t < 1$ the ellipse field corresponding to the Beltrami coefficient $t\nu$ is invariant under rotation and has major axis parallel to the direction of rotation. One can argue geometrically that any solution of the Beltrami equation increases the modulus of the annulus. This can also be seen by noting that a solution in $\zeta$-coordinates is given explicitly on the annulus by $|\zeta|^{2t/(1-t)}\zeta$.

Consider $N$ such annuli $U_1, ..., U_N$ from different cycles of Herman rings. Define $E_j$ and $\nu_j$ as above, and for $t \in \mathbf{R}^N$ satisfying $0 < t_j < 1$, $1 \le j \le N$, define the Beltrami coefficient $\mu_t$ to be $t_j \nu_j$ on $E_j$ as above. Extend $\mu_t$ to the inverse iterates of the $E_j$'s so that the corresponding ellipse field is invariant, and set $\mu_t = 0$ elsewhere. We place $\infty$ in $U_1 \backslash E_1$, and then $\mu_t$ has compact support in $\mathbf{C}$. (This was our sole purpose in introducing $E$ above.) Let $f_t$ be the normalized solution of the Beltrami equation. Then $R_t = f_t \circ R \circ f_t^{-1}$ is rational of degree $d$, and $R_t$ moves continuously with the parameter $t$. Each $f_t(U_j)$ is a Herman ring for $R_t$, for which the modulus is a strictly increasing function of $t_j$. Thus there is an open set $W$ in the positive unit cube in $\mathbf{R}^N$ such that the rational functions $R_t$, $t \in W$, are all distinct. We parametrize the rational functions by $2d+1$ complex parameters (coefficients of polynomials), and we obtain a continuous one-to-one map of $W$ into $\mathbf{C}^{2d+1} \cong \mathbf{R}^{4d+2}$. Since topological maps cannot lower dimension ([**HuW**]), we obtain $N \le 4d + 2$. $\square$

## 5.   A Quasiconformal Surgical Procedure

In some circumstances, quasiconformal surgery can be used to convert attracting fixed points to superattracting.

THEOREM 5.1. *Let $U$ be a simply connected component of the Fatou set $\mathcal{F}$ that contains an attracting fixed point, on which $R$ is $m$-to-1. Then there are a rational function $R_0$, a quasiconformal homeomorphism $\psi$ of $\overline{\mathbf{C}}$, and a compact subset $E$ of $U$, such that $\psi$ is analytic on $U \backslash E$, $\psi$ is analytic on all components of $\mathcal{F}$ not iterated to $U$, $\psi \circ R \circ \psi^{-1} = R_0$ on $\psi(U \backslash E)$, and $R_0$ on $\psi(U)$ is conjugate to $\zeta^m$ on $\Delta$.*

*Proof.* Let $\gamma_2$ be a smooth simple closed Jordan curve in $U$ surrounding the critical points of $R$, such that $\gamma_1 = R^{-1}(\gamma_2)$ is a Jordan curve between $\gamma_2$ and $\partial U$, mapped $m$-to-1 onto $\gamma_2$ by $R$. Fix $\rho > 0$. Define $h$ to be the conformal map of the inside of $\gamma_2$ onto the disk $\{|\zeta| < \rho^m\}$ sending $w_0$ to 0, and extend $h$ smoothly to map the annular region between $\gamma_1$ and $\gamma_2$ homeomorphically to the annulus $\{\rho^m < |\zeta| < \rho\}$, so that $h(z)^m = h(R(z))$ for $z \in \gamma_1$. Define $g(z)$ to be $h^{-1}(h(z)^m)$ inside $\gamma_1$, and set $g(z) = R(z)$ elsewhere in $\overline{\mathbf{C}}$. Thus $g$ combines the dynamics of $R$ outside $\gamma_1$ and of $h^{-1} \circ h^m \sim \zeta^m$ inside $\gamma_2$, as in the proof of Theorem VI.2.1.

Define an ellipse field to be circles on the part of $U$ inside $\gamma_2$, and extend it to $\overline{\mathbb{C}}$ to be $g$-invariant and to be circles outside $\cup R^{-n}(U)$. Let $\psi$ be the solution of the corresponding Beltrami equation. Then $R_0 = \psi \circ g \circ \psi^{-1}$ is a rational function, which has the same dynamics as $g$. Thus $R_0$ is $m$-to-1 on $\psi(U)$ and has a superattracting fixed point as its only critical point in $\psi(U)$, so $R_0$ is conjugate to $\zeta^m$. $\square$

Note that the Julia set of $R_0$ is the $\psi$-image of the Julia set of $R$, and the dynamics of $R_0$ outside $\psi(U)$ are the same as those of $R$ outside $U$. We are now in a position to extend Theorem V.4.3 to cover attracting fixed points.

THEOREM 5.2. *Suppose that each critical point of $R$ on $J$ is strictly preperiodic. If $U$ is a simply connected component of $\mathcal{F}$ that has an attracting periodic point, then $\partial U$ is locally connected.*

*Proof.* We may assume $U$ has a fixed point and $R$ is an $m$-to-1 branched cover of $U$ over itself. Replacing $R$ by the rational function $R_0$ of the preceding theorem, we may also assume that $R$ on $U$ is conjugate to $\zeta^m$ on $\Delta$. Then we define the curves $\gamma_n$ in $U$ exactly as in the proof of Theorem V.4.1, so that $R$ is an $m$-fold cover of $\gamma_{n+1}$ over $\gamma_n$. As in Theorem V.4.3, we have a bounded metric on the part of $U$ near $J$ for which $R$ is expanding. Now the proof of Theorem V.4.3 goes through unaltered. $\square$

THEOREM 5.3. *If there are two completely invariant components of the Fatou set, at least one of which has an attracting fixed point, then the Julia set $J$ is a simple closed Jordan curve.*

*Proof.* In this case, $J$ is adherent to each component of $\mathcal{F}$, so $J$ is connected. Since $\mathcal{F}$ contains all the critical points of $R$, Theorem 5.2 applies, and $J$ is locally connected. The Riemann map from $\Delta$ to either component of $\mathcal{F}$ extends continuously to $\overline{\Delta}$ and is easily seen to be one-to-one on $\partial \Delta$. $\square$

EXAMPLE. For $0 < b < 1$, the Julia set of $z/(z^2 - bz + 1)$ is a simple closed Jordan curve (see the example in Section V.1).

When $d = 2$, the hypothesis of the existence of an attracting fixed point in Theorem 5.3 is superfluous. In fact, if $R$ is a rational function of degree $d = 2$, then $R$ has at most one parabolic fixed point (parabolic fixed points are of multiplicity $\geq 2$). So if $\mathcal{F}$ has two completely invariant parabolic components, then they share the same fixed point, with two attracting petals, the fixed point has multiplicity three, and there can be no other fixed points. Now we claim that if $R$ has degree $d = 2$ and a single fixed point in $\overline{\mathbf{C}}$ (necessarily of multiplicity three), then $R$ is conjugate to $z/(z^2 + 1)$, and consequently (Section V.1) the Julia set $\mathcal{J}$ of $R$ is a circle in the extended plane. To see this, we place the fixed point at $z = 0$ and its other preimage at $\infty$, so we have conjugated $R$ to the form $z/(az^2 - bz + c)$. Since the fixed point is parabolic, $c = 1$. If we place a critical point at $z = 1$, we obtain $a = 1$. Since $z/(z^2 - bz + 1)$ has a fixed point at $z = b$, we have also $b = 0$.

EXAMPLE. The disguised Blaschke product $(3z^2 + 1)/(z^2 + 3)$ is conjugate to $z/(z^2 + 1)$, as it has only the one parabolic fixed point at $z = 1$.

# VII
# Local Geometry of the Fatou Set

We focus on components of the Fatou set that are simply connected, and we study the geometry and dynamics near the boundary.

## 1. Invariant Spirals

We know by Theorem III.3.1 that repelling periodic points are dense in the Julia set $\mathcal{J}$. Let $z_0$ be such a repelling periodic point, with multiplier $\lambda = (R^m)'(z_0)$. Locally $R^m(z)$ behaves like $z - z_0 \to \lambda(z - z_0)$ near $z_0$, and this seems to indicate that the Fatou set of $R$ contains a logarithmic spiral with endpoint at $z_0$. We can then conclude that $\mathcal{J}$ has a very complicated structure as soon as $\arg(R^m)'(z_0)$ is generically irrational.

We may assume that $z_0$ is a repelling fixed point, and we assume also that $z_0$ is on the boundary of an invariant component $U$ of the Fatou set $\mathcal{F}$. We wish to construct curves in $U$ terminating at $z_0$. Our strategy is as follows. Let $g$ be the branch of $R^{-1}$ fixing $z_0$, and denote $\Delta_\varepsilon = \{|z - z_0| < \varepsilon\}$. For $\varepsilon$ small, $g$ is a contraction mapping of $\Delta_\varepsilon$, with multiplier $1/\lambda$ at its attracting fixed point $z_0$. Let $w_0 \in U \cap \Delta_\varepsilon$ be near $z_0$. At issue is whether some iterate $w_n = g^n(w_0)$ lies in the same component of $U \cap \Delta_\varepsilon$ as $w_0$. If so, we join $w_0$ to $w_n$ by an arc $\gamma_0$ in $U \cap \Delta_\varepsilon$, and the union of the images of $\gamma_0$ under $g^{jn}$, $j \geq 0$, forms a curve $\gamma$ in $U \cap \Delta_\varepsilon$ terminating at $z_0$. The curve $\gamma$ is invariant under $g^n$, and $R^n(\gamma) \supset \gamma$.

Suppose there are finitely many connected components $U_1, ..., U_m$ of $U \cap \Delta_\varepsilon$ that fill out a neighborhood of $z_0$ in $U$, that is, that include all points of $U \cap \Delta_\rho$ for $\rho > 0$ sufficiently small. We may assume that each $U_j$ contains $z_0$ on its boundary. Evidently each $g(U_j)$ is either a subset of some $U_k$ or disjoint from $U$. If $z \in U_k$ is near $z_0$, then $w = R(z)$ belongs to some $U_j$, and $g(U_j) \subset U_k$. In view of the finiteness assumption, $g$ thus determines a permutation of the $U_j$'s, and consequently some iterate $g^n$ leaves invariant each $U_j$. Now the argument in the preceding paragraph shows that every $w_0 \in U \cap \Delta_\rho$ can be joined to $z_0$ by a curve $\gamma$ in $U \cap \Delta_\varepsilon$ satisfying $R^n(\gamma) \supset \gamma$.

With more effort, we establish the following.

THEOREM 1.1. *Let $U$ be a simply connected and completely invariant component of the Fatou set $\mathcal{F}$, and let $z_0 \in \partial U$ be a repelling fixed point of $R$. Then there exists $n \geq 1$ such that any $w_0 \in U$ sufficiently close to $z_0$ can be joined to $z_0$ by a curve $\gamma$ in $U$ satisfying $R^n(\gamma) \supset \gamma$.*

*Proof.* We continue with the notation introduced above. The complete invariance of $U$ guarantees that $g$ maps $U \cap \Delta_\varepsilon$ into itself. Fix $w_0 \in U \cap \Delta_\varepsilon$, and define $w_j = g^j(w_0)$.

Fix a point $b \in U$, $b \notin \Delta_\varepsilon$, and consider Green's function $G(z, b) = G(z)$. In terms of the conformal map $\psi(\zeta) = z$ of the open unit disk $\Delta$ onto $U$, $\psi(0) = b$, we have $G(z) = -\log|\psi^{-1}(z)|$. Now $B = \psi^{-1} \circ R \circ \psi$ is a $d$-sheeted covering of $\Delta$ over $\Delta$, hence a Blaschke product of degree $d$. If $\rho < 1$ is such that $B(\zeta)$ has no zeros for $\rho \leq |\zeta| \leq 1$, then, for $c > 0$ sufficiently small, $\log|\zeta| \leq c\log|B(\zeta)|$ on the boundary of the annulus and hence on the entire annulus. In terms of Green's function for $U$, this becomes

$$G(z) \geq cG(R(z)), \qquad z \in U, \ z \text{ near } \partial U.$$

Iterating this, we obtain $G(w_j) \geq c^j G(w_0)$. We also choose $c_0 > 0$ so small that $G(z, w_0) \geq c_0 G(z, b)$ on a small circle centered at $b$. By the maximum principle, the estimate persists for all $z \in U$ outside the circle, and we obtain

$$G(w_j, w_0) \geq c_0 c^j G(w_0). \qquad (1.1)$$

Now suppose the points $w_j$ all lie in different components of $U \cap \Delta_\varepsilon$. Note that $|w_j - z_0| \leq C|\lambda|^{-j}$ for some $C > 0$. Let $\gamma_j$ be the geodesic curve from $w_0$ to $w_j$, for which we have the estimate of Theorem I.4.4. Fix $N$ large. For $C|\lambda|^{-N} < r < \varepsilon$ and $j > N$, let $\delta_j(r)$ be the distance to $\partial U$ of the point where $\gamma_j$ meets the circle $\{|z - z_0| = r\}$ for the last time. The $\gamma_j$'s must pass through different arcs of $U \cap \{|z - z_0| = r\}$, or else we could join some $w_j$ to a $w_k$ by an arc in $U \cap \Delta_\varepsilon$. Thus

$$\sum_{j=N+1}^{N+p} \delta_j(r) \le 2\pi r,$$

and by the Cauchy–Schwarz inequality,

$$p^2 \le \left( \sum_{j=N+1}^{N+p} \delta_j(r) \right) \left( \sum_{j=N+1}^{N+p} \frac{1}{\delta_j(r)} \right) \le 2\pi r \sum_{j=N+1}^{N+p} \frac{1}{\delta_j(r)}.$$

Hence

$$\sum_{j=N+1}^{N+p} \int_{\gamma_j} \frac{|dz|}{\delta_j(|z|)} \ge \frac{p^2}{2\pi} \int_{C|\lambda|^{-N}}^{\varepsilon} \frac{dr}{r} \ge c_1 N p^2.$$

The integral over one of the curves $\gamma_j$ must exceed $c_1 N p$. If we take $p = N$, choose such a $\gamma_j$, and use the estimate of Theorem I.4.4, we obtain for this $j$ the estimate

$$G(w_j, w_0) \le 3 \exp(-c_1 N^2/2). \tag{1.2}$$

On the other hand, from (1.1) and $j \le 2N$ we have $G(w_j, w_0) \ge c_0 c^{2N} G(w_0)$, and this contradicts (1.2) for $N$ large.

Hence for some $N < j < j + n \le 2N$, $w_j$ can be joined to $w_{j+n}$ by an arc in $U \cap \Delta_\varepsilon$. The union of the images of the arc under iterates of $g^n$ forms a curve in $U \cap \Delta_\varepsilon$ starting at $w_j$ and terminating at $z_0$. The image $\gamma$ of this curve under $R^j$ starts at $w_0$, terminates at $z_0$, and satisfies $R^n(\gamma) \supset \gamma$.

It remains to show that $n$ can be chosen independent of $w_0$, and this part of the argument is topological. Replacing $R$ by an iterate, we can assume there is already a path $\beta$ in $U \cap \Delta_\varepsilon$ terminating at $z_0$ and satisfying $g(\beta) \subset \beta$. We repeat the above discussion, and we want to show that $w_j$ and $w_{j+1}$ lie in the same component of $U \cap \Delta_\varepsilon$ for some $j$, since then we obtain $\gamma$ satisfying $R(\gamma) \supset \gamma$. We assume that the points $w_{j+k}$ are in different components of $U \cap \Delta_\varepsilon$ for $0 \le k < n$. We can also assume that these are different from the component containing $\beta$, which is invariant under $g$. Let $\beta_0$ denote the path constructed above, from $w_j$ through $w_{j+n}$ to $z_0$, and let

$\beta_k = g^k(\beta)$. The paths $\beta, \beta_0, ..., \beta_{n-1}$ have a circular ordering, which is the ordering of the last points where the paths cross a fixed small circle $\{|z - z_0| = r\}$. This ordering is independent of $r$, and it is preserved by $g$. Since the curve $\beta$ is fixed by $g$, and since the others are permuted, in fact the others must be fixed also. Thus $w_{j+1} = g(w_j)$ lies in the same component as $w_j$, as required. $\square$

Simple examples show we cannot take $n = 1$ in Theorem 1.1. Consider $R(z) = z^2 - 2$, which has Julia set $\mathcal{J} = [-2, 2]$. The repelling fixed point $-1$, with multiplier $-2$, is accessible from each side of $\mathcal{J}$ by paths in $\mathcal{F}$ satisfying $R^2(\gamma) \supset \gamma$ but by no path satisfying $R(\gamma) \supset \gamma$. However, there is an analogous result for rationally neutral fixed points, and here we *can* take $n = 1$.

THEOREM 1.2. *Let $z_0$ be a parabolic fixed point of $R$ with multiplier $R'(z_0) = 1$. Let $U$ be a simply connected and completely invariant component of $\mathcal{F}$ with $z_0 \in \partial U$ and $U$ not an attracting petal for $z_0$. Then if $w_0 \in U$ is sufficiently close to $z_0$, there exists a curve $\gamma \subset U$ joining $z_0$ to $w_0$, such that $R(\gamma) \supset \gamma$.*

*Proof.* The proof of Theorem 1.1 carries through with minor modifications. In this case, $U$ is included in a cusp at $z_0$, and the discussion of the Fatou coordinate, applied to $g$, shows for $\varepsilon$ small that $g$ maps $U \cap \Delta_\varepsilon$ into itself. Moreover, $|w_k - z_0| \sim 1/k^{1/m}$, where $m$ is the number of petals at $z_0$. This time $\gamma_j$ satisfies

$$\int_{\gamma_j} \frac{|dz|}{\delta_j(|z|)} \geq \frac{N}{2\pi} \int_{CN^{-1/m}}^{\varepsilon} \frac{dr}{r} \geq c_2 N \log N + c_3 N + c_4,$$

where $c_2 > 0$. As before, the estimate of Theorem I.4.4 leads to a contradiction for large $N$.

Again we obtain a curve $\gamma$ in $U \cap \Delta_\varepsilon$ joining $w_j$ to $w_k$ for some $k > j$. Consider the Fatou coordinate function $\varphi$ for $g$, which conjugates $g$ to translation by 1. The curve $\varphi(\gamma)$ goes from $\varphi(w_j)$ to $\varphi(w_j) + k - j$, while $\varphi(g(\gamma))$ is the translate of $\varphi(\gamma)$ by 1. If $k - j \geq 2$, the curve must intersect its translate, and we find a curve from $w_j$ to $w_{j+1}$ in $U \cap \Delta_\varepsilon$. As before the various iterates generate a curve from $w_0$ to $z_0$, this time satisfying $R(\gamma) \supset \gamma$. $\square$

Theorems 1.1 and 1.2 were proved in [Ca2]. The proof methods are related to those of the Ahlfors–Carleman–Denjoy theorem (see [**Beu**], [**Fu**]). A.E. Eremenko and G.M. Levin (1989) have obtained invariant curves in $U$ terminating at repelling fixed points, without the hypothesis of simple connectivity. They treat polynomials, with $U = A(\infty)$, and they apply the subharmonic version of the Ahlfors–Carleman–Denjoy theorem to the composition of the inverse of Koenigs' coordinate function (an entire function of finite order; see Section II.3) and Green's function.

## 2.   Repelling Arms

Let $U$ be an invariant, simply connected component of the Fatou set $\mathcal{F}$, and let $z_0 \in \partial U$ be a fixed point of $R(z)$. Let $V$ be an open subset of $U$, and suppose there is a conformal map $\zeta = \varphi(z)$ of $V$ into a horizontal strip $S = \{a < \operatorname{Im}\zeta < b\}$ whose image includes a left half-strip $\{\zeta \in S : \operatorname{Re}\zeta < -C\}$, such that $z \to z_0$ as $\operatorname{Re}\zeta \to -\infty$ with $\operatorname{Im}\zeta$ fixed; $z \to \mathcal{J}$ as $\operatorname{Im}\zeta \to a$ or $\operatorname{Im}\zeta \to b$; and $\varphi$ conjugates $R$ to the translation $\zeta \to \zeta + 1$,

$$\varphi(R(z)) = \varphi(z) + 1, \qquad z \in V, \ R(z) \in V.$$

In this situation, we say that $V$ is a *repelling arm of $\mathcal{F}$ terminating at $z_0$*. The quotient space $V/R$ is conformally equivalent to an annulus with modulus $b - a$, so that the height of the strip is uniquely determined. The coordinate $\varphi$ is unique, up to a translation. Applying Lemma IV.2.4 to $R^{-1}$, we see that the fixed point $z_0$ is either a repelling fixed point or a parabolic fixed point with multiplier 1. If $z_0$ is parabolic, then $V$ approaches $z_0$ through a gap between consecutive attracting petals.

A *periodic repelling arm* terminating at a periodic point is defined in the obvious way. The topological argument at the end of the proof of Theorem 1.1 shows that periodic repelling arms terminating at the same point have the same period. Moreover, if the fixed point is parabolic, and if a cycle of periodic repelling arms terminates at the fixed point through the same cusp, then in fact the cycle consists of only one repelling arm.

EXAMPLE. Let $B(z)$ be a finite Blaschke product of degree $d \geq 2$ whose Julia set coincides with the entire unit circle $\partial\Delta$, and suppose $z_0 \in \partial\Delta$ is a repelling fixed point of $B$. Let $\tau$ be Koenigs' coordinate function at $z_0$, so that $\tau(z_0) = 0$, and $\tau(B(z)) = \lambda\tau(z)$ in a neighborhood of $z_0$. Since $\arg B(z)$ is increasing around $\partial\Delta$, the multiplier $\lambda$ is positive, $\lambda > 1$. Since $B$ leaves $\partial\Delta$ invariant, we can assume $\tau$ maps $\Delta$ to the lower half-plane. Then

$$\varphi(z) = \frac{\operatorname{Log}\tau(z)}{\log\lambda} \qquad \text{(principal branch)}$$

coordinatizes two repelling arms terminating at $z_0$, one in $\Delta$ and the other in $\mathbf{C}\backslash\overline{\Delta}$, each mapped to a half-strip of height $\pi/\log\lambda$.

THEOREM 2.1. *Suppose $U$ is an invariant component of $\mathcal{F}$ that is simply connected, let $\psi : \Delta \to U$ be the Riemann mapping, and set $B = \psi^{-1} \circ R \circ \psi$, a finite Blaschke product. Suppose $\zeta_0 \in \partial\Delta$ is a fixed point for $B(\zeta)$. Then $\psi(\zeta)$ has a nontangential limit $z_0$ at $\zeta_0$, and $z_0$ is a fixed point for $R(z)$ that is either repelling or parabolic with multiplier 1. If $\zeta_0$ is a repelling fixed point for $B(\zeta)$, then for $\rho > 0$ sufficiently small, $\psi(\{\Delta \cap |\zeta - \zeta_0| < \rho\})$ is a repelling arm for $R(z)$. Otherwise, $\zeta_0$ is a parabolic fixed point for $B(\zeta)$ with Julia set $\partial\Delta$, and $z_0$ is a parabolic fixed point for $R(z)$ which has $U$ as an attracting petal.*

*Proof.* Since $B$ is proper, it is a finite Blaschke product. Assume $\zeta_0 = 1$ is fixed by $B$. Let $r_n$ be any sequence of radii increasing to 1 such that $\psi(r_n)$ converges, say to $z_0 \neq \infty$. Now the conformal self-map $(\zeta - r_n)/(1 - r_n\zeta)$ of $\Delta$ maps $r_n$ to 0 and $B(r_n)$ to

$$\frac{B(r_n) - r_n}{1 - r_n B(r_n)} = -\frac{[(B(r_n) - 1)/(r_n - 1)] - 1}{r_n[(B(r_n) - 1)/(r_n - 1)] + 1},$$

whose modulus tends to

$$\left|\frac{B'(1) - 1}{B'(1) + 1}\right| < 1.$$

Thus the hyperbolic distance in $\Delta$ from $r_n$ to $B(r_n)$ is uniformly bounded, and the hyperbolic distance in $U$ from $\psi(r_n)$ to $\psi(B(r_n)) = R(\psi(r_n))$ is also bounded. The comparison of the hyperbolic and euclidean metrics in Theorem I.4.3 shows that $|R(\psi(r_n)) - \psi(r_n)| \to 0$. In the limit we obtain $R(z_0) = z_0$. Since $R$ has only a finite number of fixed points, $\psi(r)$ must accumulate at $z_0$ as $r$ increases to 1, and $\psi(\zeta)$ has nontangential limit $z_0$ at 1.

If the fixed point $\zeta_0 \in \partial\Delta$ is repelling, we compose $\psi^{-1}$ with the logarithm of Koenigs' coordinate function, as in the example above, to obtain a repelling arm at $z_0$. This implies $z_0$ is either repelling or parabolic with multiplier 1.

Suppose finally that $\zeta_0$ is not repelling. Then $B^n(\zeta) \to \zeta_0$ on $\Delta$, so that $R^n(z) \to \partial U$ for $z \in U$. By Lemma IV.2.4, $R^n(z)$ converges on $U$ to a parabolic fixed point $z_1$ of $R(z)$ with multiplier 1. By joining an appropriate $\zeta \in \Delta$ to $B(\zeta)$ and iterating, we obtain a curve in $\Delta$ terminating at $\zeta_0$ along which $\psi(\zeta)$ has limit $z_1$. Lindelöf's theorem implies $z_1 = z_0$. Thus in this case $z_0$ is a parabolic fixed point and $U$ is an attracting petal at $z_0$.

Consider two arcs $\gamma_+$ and $\gamma_-$ in $U$ starting at $z_0$ and terminating at points $z_+, z_- \in \mathcal{J}$ which lie near $z_0$ in the respective repelling cusps on each side of the attracting petal at $z_0$. These curves divide $U$ into three pieces, one containing the attracting direction, the other two $U_\pm$ lying between $\gamma_\pm$ and the part of $\mathcal{J}$ in the repelling cusps. In view of our analysis of parabolic fixed points, we can choose $\gamma_+$ and $\gamma_-$ so that $U_+$ and $U_-$ are invariant under $g$. Let $\zeta_\pm$ be the terminal points on $\partial \Delta$ of the curves $\psi^{-1}(\gamma_\pm)$ as $z \to z_\pm$. The domains $\psi^{-1}(U_\pm)$ are contiguous to the two arcs between $\zeta_0$ and $\zeta_\pm$, and these arcs must be on different sides of $\zeta_0$. The curve $\psi^{-1}(g^n(\gamma_+))$ in $U_+$ terminates at $\zeta_0$ and another point $\zeta_n$. Since $g^n(U_+)$ decreases, $\zeta_n$ moves monotonically towards $\zeta_0$, hence converges to a point $\zeta^*$. Since $g^n(z_+)$ moves towards $z_0$ in the repelling cusp, one sees that any nontangential limit of $\psi$ on the arc between $\zeta^*$ and $\zeta_0$ must coincide with $z_0$, hence $\zeta^* = \zeta_0$. Evidently $B(\zeta_{n+1}) = \zeta_n$, so the direction along the arc from $\zeta_0$ to $\zeta_+$ is repelling for $B$. Similarly, the direction from $\zeta_0$ to $\zeta_-$ is repelling, so $\zeta_0$ is a parabolic fixed point for $B$ with two petals.  $\square$

EXAMPLE. Consider $R(z) = z^2 + 1/4$. The Julia set $\mathcal{J}$ of $R$ is the cauliflower set depicted in Figure 5 of Section VIII.1, which was shown in Section V.4 to be a simple closed Jordan curve. On the bounded component of the Fatou set $\mathcal{F}$, $R$ is conjugate to a Blaschke product on $\Delta$ of degree 2, with one parabolic fixed point on $\partial \Delta$ corresponding to the parabolic fixed point of $R$ at $1/2$. According to the discussion in Section VI.5, $R$ is in fact conjugate to $B(\zeta) = (3\zeta^2 + 1)/(3 + \zeta^2)$. The unbounded component $A(\infty)$ of $\mathcal{F}$ has a repelling arm at the parabolic fixed point $1/2$, which issues from $1/2$ along the real axis in the positive direction. The action of $R$ on $A(\infty)$ is conjugate to the Blaschke product $\zeta^2$, which has a superattracting fixed point at 0 (corresponding to $\infty$), and a repelling fixed point at $\zeta = 1$ (corresponding to the parabolic fixed point at $z = 1/2$).

The argument in Theorem 2.1 is reversible, as follows easily from Lindelöf's theorem. A periodic repelling arm in an invariant component $U$ arises from a unique repelling periodic point on $\partial\Delta$, and the period of the arm coincides with the period of the periodic point. In the case where $\partial U$ is locally connected, we can say more.

THEOREM 2.2. *Suppose $U$ is an invariant component of the Fatou set $\mathcal{F}$ that is simply connected, and suppose $\partial U$ is locally connected. Let $\psi : \overline{\Delta} \to \overline{U}$ be the Riemann mapping, with associated Blaschke product $B = \psi^{-1} \circ R \circ \psi$. Suppose $z_0 \in \partial U$ is a repelling periodic point for $R$, or that $z_0$ is a parabolic periodic point and $U$ is not an attracting petal for $z_0$. Then $\psi^{-1}(z_0) = \{\zeta_1, ..., \zeta_N\}$ is finite. Each $\zeta_j$ is a repelling periodic point for $B$ and determines a periodic repelling arm in $U$ that terminates at $z_0$. These $N$ periodic repelling arms fill out a neighborhood of $z_0$ in $U$.*

*Proof.* We assume $z_0$ is a repelling fixed point (the parabolic case is almost the same, and there is also an analogous result if $U$ is an attracting petal for $z_0$). We use the notation of Section 1, and we take $\delta > 0$ so small that $|\psi(\zeta) - \psi(\xi)| < \varepsilon/2$ for $\zeta, \xi \in \Delta$ satisfying $|\zeta - \xi| \le \delta$. If $z = \psi(\zeta) \in U \cap \Delta_{\varepsilon/2}$, then the image under $\psi$ of $\Delta \cap \Delta(\zeta, \delta)$ is contained in $U \cap \Delta_\varepsilon$. Since only finitely many of these sets $\Delta \cap \Delta(\zeta, \delta)$ can fit disjointly into $\Delta$, there are only finitely many components $U_1, ..., U_N$ of $U \cap \Delta_\varepsilon$ that contain $z_0$ in their boundary, and these fill out a neighborhood of $z_0$ in $U$. By the discussion preceding Theorem 1.1, there is in each $U_j$ a curve $\gamma_j$ invariant under some iterate of $g$ and terminating at $z_0$. Then $\psi^{-1}(\gamma_j)$ terminates at a point $\zeta_j \in \partial\Delta$, and if $R^n(\gamma_j) \supset \gamma_j$, then $\zeta_j$ is a fixed point of $B^n$. By our hypothesis, $\zeta_j$ cannot be a parabolic fixed point for $B$, so $\zeta_j$ is a repelling periodic point. It determines a repelling arm terminating at $z_0$, via Theorem 2.1, through which $\gamma_j$ approaches $z_0$. The curve $\gamma_j$ can be chosen so that its initial arc passes through any two prescribed points of $U_j$. Consequently $\psi^{-1}$ has the same limit along the iterates of these two points, and for each $w \in U_j$, $\psi^{-1}(g^{kn}(w))$ tends to $\zeta_j$ as $k \to \infty$.

Let $\rho > 0$ be small, and let $V_j = \psi(\Delta(\zeta_j, \rho) \cap \Delta)$. Then $V_j$ is a repelling arm for some $R^n$, which is contained in $U_j$ and which terminates at $z_0$. We claim that $V_1, ..., V_N$ are the only repelling arms in $U$ terminating at $z_0$. Indeed, if $V$ is such a repelling arm, say for $R^n$, then the iterates of $w \in V$ under $g^n$ eventually enter some $U_j$, and since $\psi^{-1}(g^{kn}(w))$ tends to $\zeta_j$, the arm $V$ is the same arm as $V_j$.

Suppose $\xi_0 \in \psi^{-1}(z_0)$ is distinct from the $\zeta_j$'s. We may assume that $\rho$ is less than the distance from $\xi_0$ to the $V_j$'s, and that there is $r > 0$ small such that $|\psi(\zeta) - z_0| > r$ on the circular arc $\{|\zeta - \zeta_0| = \rho\} \cap \Delta$. Let $W$ be the component of $U \cap \Delta_r$ containing the image under $\psi$ of the end of the radial segment terminating at $\xi_0$. By what has been established, with $\varepsilon$ replaced by $r$, $W$ contains a repelling arm terminating at $z_0$. However, this contradicts the fact that $W$ is disjoint from the $V_j$'s. We conclude that $\psi^{-1}(z_0) = \{\zeta_1, ..., \zeta_N\}$, and it follows easily that the $V_j$'s fill out a neighborhood of $z_0$ in $U$. $\square$

For a color illustration of such a configuration (in the Mandelbrot set), with 29 arms, see Map 38 on page 82 of [**PeR**].

## 3.  John Domains

A domain $D$ is called a *John domain* if there exists $c > 0$ such that for any $z_0 \in D$, there is an arc $\gamma$ joining $z_0$ to some fixed reference point $w_0 \in D$ satisfying

$$\mathrm{dist}(z, \partial D) \geq c|z - z_0|, \qquad z \in \gamma. \qquad (3.1)$$

If $\infty \in \partial D$, we use the spherical metric to measure the distance.

If two smooth boundary arcs meet at a corner with a positive angle, the condition (3.1) is satisfied near the vertex. If however the two boundary arcs meet tangentially at an outward-pointing cusp, the condition (3.1) cannot be satisfied. Similarly, if $D$ is a simply connected component of the Fatou set that has a repelling arm terminating at a parabolic point, then $D$ is not a John domain.

A simply connected John domain has a locally connected boundary. This can be seen easily from the proof of Carathéodory's theorem given in Section I.2. The crosscuts $\Gamma_n$ constructed there have lengths tending to 0, and the condition (3.1) then shows that the diameters of the pieces of $D$ outside the crosscuts tend to 0. This implies that the Riemann map from $\Delta$ to $D$ extends continuously to $\partial \Delta$.

The image of a John domain under a quasiconformal homeomorphism of $\overline{\mathbf{C}}$ is evidently a John domain. Thus the two complementary components of a quasicircle are John domains. It can be shown, conversely, that if the two complementary components of a simple closed Jordan curve are John domains, then the curve is a quasicircle. Thus John domains can be regarded as "one-sided quasidisks." For this and further background material on John domains, see [NaV]. Towards improving upon Theorem VI.2.1, we prove the following.

THEOREM 3.1. *Suppose $R$ is subhyperbolic on the Julia set $\mathcal{J}$. Then any simply connected component of the Fatou set $\mathcal{F}$ is a John domain.*

We carry out the proof first in the hyperbolic case, where the ideas are clearer, and then we indicate how to adapt the proof to the subhyperbolic case.

*Proof (hyperbolic case).* Let $R$ be hyperbolic, and suppose $U$ is a simply connected component of $\mathcal{F}$. Each component of $\mathcal{F}$ is iterated eventually to an attracting cycle of components. Replacing $R$ by an iterate, we may assume that $R(U)$ contains an attracting fixed point. Since $R$ is proper on $U$, $R(U)$ is also simply connected. Since $R$ is conformal on $\partial U$, it suffices to prove that $R(U)$ is a John domain. Thus we may assume that $U$ itself has an attracting fixed point. Replacing $R$ by the rational function $R_0$ of Theorem VI.5.1, which is also hyperbolic, we may further assume that $R$ on $U$ is conjugate to $\zeta^m$ on $\Delta$.

Let $\psi : \Delta \to U$ conjugate $\zeta^m$ to $R(z)$. By Theorem VI.4.2 (or by the proof of Theorem V.4.3), $\psi$ extends continuously to $\partial\Delta$. We will show that the geodesics $\{\psi(re^{i\theta}) : 0 \le r < 1\}$ can be taken to be the curves in the definition of a John domain. (This may seem fortuitous, but it turns out that in any simply connected John domain, we can take the curves in the definition to be geodesics.) Note that $R$ maps rays to rays.

Choose $\rho > 0$ so that all the inverse iterates $R^{-k}$ are almost affine on any disk of radius $\rho$ centered in a neighborhood of $\mathcal{J}$, that is, so that the estimate used in the proof of Theorem V.2.2 holds. Set $V = \psi(\{1 - \delta < |\zeta| < 1\})$, where $\delta > 0$ is so small that the affine mapping estimate holds on disks centered in $V$ of radius $\rho$. Since $\psi$ is uniformly continuous, we can assume that any two points of $V$ on the same conformal ray are within distance $\rho$ of each other. Set

$$c_0 = \inf\{\mathrm{dist}(z, \mathcal{J}) : z \in V, R(z) \notin V\}.$$

Then $c_0 < \rho$.

Suppose $z_1$ lies on the ray between $w_0$ and $z_0$. It suffices to check the estimate (3.1) for $z_0$ and $z_1$ in $V$. Let $n \geq 1$ satisfy $R^n(z_1) \in V$ and $R^{n+1}(z_1) \notin V$. Then the disk centered at $R^n(z_1)$ of radius $c_0$ is disjoint from $\mathcal{J}$. By the affine mapping property, the image of this disk under the appropriate branch of $R^{-n}$ covers a disk centered at $z_1$ of radius say half of $c_0/|(R^n)'(z_1)|$. Hence

$$\mathrm{dist}(z_1, \mathcal{J}) \geq \frac{1}{2} \frac{c_0}{|(R^n)'(z_1)|}. \tag{3.2}$$

On the other hand, since $R^n(z_0)$ and $R^n(z_1)$ lie on the same ray, they are within distance $\rho$ of each other, and the affine mapping estimate yields $|z_0 - z_1| \leq 2\rho/|(R^n)'(z_1)|$. Combined with the estimate (3.2) above, this gives (3.1), and the theorem is proved when $R$ is hyperbolic. $\square$

To carry over the proof to the subhyperbolic case, we require a version of the affine mapping estimate. For this, it is natural to work with an admissible metric, and we use the specific metric $\sigma$ constructed in the proof of Theorem V.3.1, as modified in the remark after the proof. Thus $\sigma(z)$ is a constant multiple of $|z - a_j|^{-\beta_j}$ near each of its singularities $a_j$, where $0 < \beta_j < 1$, and elsewhere $\sigma(z)$ is smooth.

Let $D_\sigma$ denote differentiation with respect to the $\sigma$-metric, defined by

$$(D_\sigma R)(z) = \frac{\sigma(R(z))}{\sigma(z)} R'(z).$$

Thus $R$ dilates infinitesimal distances at $z$ in the $\sigma$-metric by a factor of $|(D_\sigma R)(z)|$. The chain rule is valid,

$$(D_\sigma R^n)(z) = \prod_{k=0}^{n-1} (D_\sigma R)(R^k(z)),$$

as is the inverse function rule, $(D_\sigma R^{-1})(R(z)) = 1/(D_\sigma R)(z)$. From the precise form of $\sigma$ at its singularities, one checks that $|D_\sigma R|$ is smooth in a neighborhood of $\mathcal{J}$. The affine mapping estimate is now as follows.

LEMMA 3.2. *Suppose $R$ is subhyperbolic on $\mathcal{J}$. Let $\sigma$ be the admissible metric introduced above. Then there is a neighborhood $V$ of $\mathcal{J}$ such that*

$$\frac{(D_\sigma R^{-n})(w)}{(D_\sigma R^{-n})(z)} = 1 + d_\sigma(z,w)\mathcal{O}(1), \qquad z, w \in V, n \geq 1.$$

*Proof.* Here we can specify $R^{-n}(z)$ to be an arbitrary branch, and then the determination of $R^{-n}(w)$ is that obtained by continuing $R^{-n}(z)$ along the shortest path between $z$ and $w$ in the $\sigma$-metric (irrespective of singularities of $R^{-n}$). Since $R$ is strictly expanding with respect to $\sigma$, the inverse branches satisfy $|(D_\sigma R^{-1})(z)| \leq c < 1$ in a neighborhood of $\mathcal{J}$. The neighborhood can be chosen to be invariant under the inverse branches. By integrating, we obtain

$$d_\sigma(R^{-k}(z), R^{-k}(w)) \leq c^k d_\sigma(z,w)$$

whenever the shortest path between $z$ and $w$ lies in the neighborhood of $\mathcal{J}$ and in particular whenever $z$ is near $w$. Thus

$$\left| 1 - \frac{(D_\sigma R)(R^{-k})(z))}{(D_\sigma R)(R^{-k})(w))} \right| = \left| \frac{(D_\sigma R)(R^{-k})(z)) - (D_\sigma R)(R^{-k})(w))}{(D_\sigma R)(R^{-k})(w))} \right|$$
$$\leq C_1 |(R^{-k})(z) - (R^{-k})(w)|$$
$$\leq C_2 d_\sigma(R^{-k}(z), R^{-k}(w)) \leq C_2 c^k d_\sigma(z,w).$$

Using the chain rule as in the proof of Theorem V.2.3, we obtain

$$\frac{(D_\sigma R^{-n})(w)}{(D_\sigma R^{-n})(z)} = \prod_{k=1}^{n} [1 + c^k d_\sigma(z,w)\mathcal{O}(1)],$$

which yields the desired estimate. $\square$

*Proof of Theorem 3.1 (subhyperbolic case).* We aim to show that $U$ satisfies the John condition (3.1), with distances measured in the $\sigma$-metric rather than the euclidean metric. From this, it is straightforward to conclude that $U$ is a John domain, and the details are left as an exercise. (Note that a problem arises in comparing the metrics only at singularities, so one can focus on $\sigma(z) = |z|^{-\beta}$ inside the unit disk.)

Since $|D_\sigma(R)|$ is smooth, bounded above and away from zero near $\mathcal{J}$, the $\sigma$-geometry of $U$ near the boundary is the same as that of

$R(U)$, so we may proceed as before and assume that $R(z)$ on $U$ is conjugate to $\zeta^m$ on $\Delta$. Again the conjugation extends continuously to $\partial U$, hence is uniformly continuous with respect to the $\sigma$-metric.

Choose $\rho > 0$ as before, with $R^{-k}$ almost affine on $\sigma$-disks of radius $\rho$, and choose $V$ as before. Define $c_0$ in terms of the $\sigma$-metric, so that if $z \in V$ but $R(z) \notin V$, then the $\sigma$-distance from $z$ to $\mathcal{J}$ is at least $c_0$. For $z_1$ on the ray between $w_0$ and $z_0$, we obtain this time an estimate of the form

$$\mathrm{dist}_\sigma(z_1, \mathcal{J}) \geq \frac{1}{2}\frac{c_0}{|(D_\sigma R^n)(z_1)|}. \tag{3.3}$$

Since $R^n(z_0)$ and $R^n(z_1)$ lie on the same ray, they are contained in a $\sigma$-disk of radius $\rho$, as is the ray connecting them. We may assume that this disk also contains the shortest path between the points. If there are no singular points $a_j$ in this disk, the analytic branch of $R^{-n}$ sending $R^n(z_1)$ to $z_1$ also sends $R^n(z_0)$ to $z_0$. This yields an estimate of the form

$$d_\sigma(z_0, z_1) \leq \frac{4\rho}{|(D_\sigma R^n)(z_1)|}. \tag{3.4}$$

If there is a singular point $a_j$ in this disk, we connect $R^n(z_1)$ to $R^n(z_0)$ by a path in the disk of $\sigma$-length at most $3\rho$, such that the continuation of the branch of $R^{-n}$ mapping $R^n(z_1)$ to $z_1$ along the path sends $R^n(z_0)$ to $z_0$. By integrating along this path, we obtain again an estimate of the form (3.4). Now (3.3) and (3.4) show that $U$ is a John domain in the $\sigma$-metric, as required. $\square$

Suppose $P(z)$ is a subhyperbolic polynomial with a connected Julia set $\mathcal{J}$. We claim that the boundary of each bounded component $U$ of the Fatou set $\mathcal{F}$ is a quasicircle. Indeed, $\partial U$ is locally connected, so the Riemann map from $\Delta$ to $U$ extends continuously to $\partial \Delta$. Since $\partial U$ is included in the boundary of a single component $A(\infty)$ of its complement, the Riemann map is one-to-one on $\partial \Delta$, and $\partial U$ is a simple closed Jordan curve. Now $A(\infty)$ is also a simply connected John domain. From the condition (3.1) for $A(\infty)$, one deduces easily that (3.1) holds for the larger domain $\overline{\mathbf{C}} \backslash \overline{U}$. If we invoke now the characterization of quasicircles cited earlier (but not proved), we deduce that $\partial U$ is a quasicircle. In the case of hyperbolic polynomials, this result is subsumed by a theorem of M.V. Yakobson (1984).

EXAMPLE. Each component of the body of Douady's rabbit (Figure 4 in Section VIII.1) is bounded by a quasicircle.

Theorem 3.1 and the application come from [CaJ]. Under certain hypotheses, Theorem 3.1 remains valid for parabolic components of $\mathcal{F}$ (see [CaJY]). Herman has proved (unpublished) that if the multiplier of a Siegel disk associated with a quadratic polynomial has the form $\lambda = e^{2\pi i\theta}$ where $\theta$ satisfies the Diophantine condition (6.2) of Section II.6 for some exponent $\mu$, then the boundary of the Siegel disk is a quasicircle if and only if the Diophantine condition is satisfied with $\mu = 2$. In the case $\mu = 2$, C. Petersen has recently proved (to appear in *Acta Math.*) that the filled-in Julia set is locally connected. These results settle a conjecture discussed in [Do3].

# VIII
# Quadratic Polynomials

We are interested in the dependence of a dynamical system on parameters, and we focus on the maps $z \to z^2 + c$. This leads to the Mandelbrot set $\mathcal{M}$ in parameter space, which has a universal character in the sense that similar-looking sets show up when one studies very general parameter dependence. One striking feature of $\mathcal{M}$ is that shapes of certain of the Julia sets $\mathcal{J}_c$ in dynamic space ($z$-space) are reflected in the shape of $\mathcal{M}$ near the corresponding points in parameter space ($c$-space).

## 1. The Mandelbrot Set

A quadratic polynomial $P(z)$ can be conjugated by $z' = az$ to a monic polynomial $z^2 + \alpha z + \beta$. This can be further conjugated by a translation $z' = z + b$ to move any given point to 0. If we move one of the fixed points to 0 we have conjugated $P$ to the form $\lambda z + z^2$, where $\lambda$ is the multiplier of the fixed point. This does not determine the conjugacy class uniquely, as we can place the second fixed point at 0. If we move the critical point to 0 we have conjugated $P$ to the form

$$P_c(z) = z^2 + c,$$

and now it is easy to check that different $c$'s correspond to different conjugacy classes. Thus we can regard the $c$-plane as representing conjugacy classes of quadratic polynomials (that is, it is a "moduli space"). We are interested in how the dynamic behavior of $P_c$ depends on the parameter $c$. The following dichotomy is an immediate consequence of Theorems III.4.1 and III.4.2.

THEOREM 1.1. *If* $P_c^n(0) \to \infty$, *then the Julia set* $\mathcal{J}_c$ *is totally disconnected. Otherwise* $P_c^n(0)$ *is bounded, and the Julia set is connected.*

The set of parameter values $c$ such that $P_c^n(0)$ is bounded is of special interest. It is called the *Mandelbrot set* and denoted by $\mathcal{M}$. Thus $c \in \mathcal{M}$ if and only if $0$ does not belong to the basin of attraction of the superattracting fixed point at $\infty$.

THEOREM 1.2. *The Mandelbrot set* $\mathcal{M}$ *is a closed simply connected subset of the disk* $\{|c| \leq 2\}$, *which meets the real axis in the interval* $[-2, 1/4]$. *Moreover,* $\mathcal{M}$ *consists of precisely those* $c$ *such that* $|P_c^n(0)| \leq 2$ *for all* $n \geq 1$.

*Proof.* If $|c| > 2$, one shows by induction that

$$|P_c^n(0)| \geq |c|(|c| - 1)^{2^{n-1}}, \qquad n \geq 1,$$

so $|P_c^n(0)| \to \infty$ and $c \notin \mathcal{M}$. Thus $|c| \leq 2$ for $c \in \mathcal{M}$.

Suppose $|P_c^m(0)| = 2 + \delta > 2$ for some $m \geq 1$. If $|c| = |P_c(0)| > 2$, then $c \notin \mathcal{M}$. If $|c| \leq 2$, then $|P_c^{m+1}(0)| \geq (2 + \delta)^2 - 2 \geq 2 + 4\delta$. Proceeding by induction, we obtain $|P_c^{m+k}(0)| \geq 2 + 4^k\delta \to \infty$, and again $c \notin \mathcal{M}$. This proves the final statement of the theorem, from which it follows that $\mathcal{M}$ is closed. By the maximum principle, $\mathbf{C}\backslash\mathcal{M}$ has no bounded components, so $\mathbf{C}\backslash\mathcal{M}$ is connected, and $\mathcal{M}$ is simply connected.

If $c$ is real, then $P_c(x) - x$ has no real roots if $c > 1/4$, one root at $1/2$ if $c = 1/4$, and two real roots if $c < 1/4$. If $c > 1/4$, $P_c^n(0)$ is increasing and must go to infinity, since any finite limit point would satisfy $P_c(x) = x$. If $c \leq 1/4$, let $a = (1 + \sqrt{1 - 4c})/2$ be the larger root of $P_c(x) - x$. If additionally $c \geq -2$, one checks that $a \geq |c| = |P_c(0)|$. Then $|P_c^n(0)| \leq a$ implies $|P_c^{n+1}(0)| = |P_c^n(0)^2 + c| \leq a^2 + c = a$, and the sequence is bounded. Thus $\mathcal{M} \cap \mathbf{R} = [-2, 1/4]$. $\square$

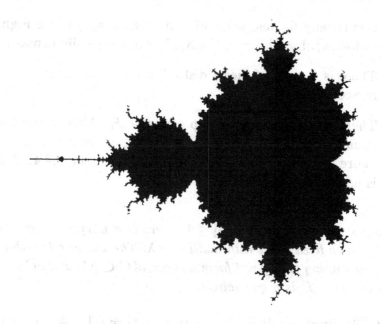

FIGURE 1. The Mandelbrot set $\mathcal{M}$.

The theorem suggests a simple algorithm for computing $\mathcal{M}$. If $|P_c^n(0)| \leq 4$ for $1 \leq n \leq 1000$, color the pixel $c$ black, otherwise color it white. See Figure 1.

We review the various possibilities for the Julia set $\mathcal{J}_c$ of $P_c$. By Douady's Theorem VI.1.2, there is at most one periodic cycle of bounded components of $\mathcal{F}_c$. By Sullivan's Theorem IV.1.3, every bounded component of $\mathcal{F}_c$ is eventually iterated into the cycle. Since Herman rings do not arise for polynomials, the classification theorem of Section IV.2 gives the following four possibilities for $c \in \mathcal{M}$.

1. There is an attracting cycle for $P_c$. Either there is an attracting fixed point, in which case there is only one bounded component of $\mathcal{F}_c$. Or the cycle has length two or more, in which case there are infinitely many bounded components of $\mathcal{F}_c$. By Theorem V.4.1, $\mathcal{J}_c$ is locally connected.

2. There is a parabolic cycle for $P_c$. Either there is a parabolic fixed point with multiplier 1, in which case there is only one bounded component of $\mathcal{F}_c$. Or the cycle of parabolic components of $\mathcal{F}_c$ has length two or more, in which case there are infinitely many bounded components of $\mathcal{F}_c$. The former case

occurs only for one value of $c$, namely, $c = 1/4$ (see Figure 5 to follow). By Theorem V.4.3, $\mathcal{J}_c$ is again locally connected.

3. There is a cycle of Siegel disks. This case has been discussed in Section VI.1.

4. There are no bounded components of $\mathcal{F}_c$. This occurs for instance for $c = -2$, where $\mathcal{J}_c = [-2, 2]$, or for $c = i$, as in the figure in Section V.4. It may happen (Theorem V.4.4) that $\mathcal{J}_c$ is not locally connected.

THEOREM 1.3. *For each $\lambda$, $|\lambda| \leq 1$, there is a unique $c = c(\lambda)$ such that $P_c$ has a fixed point with multiplier $\lambda$. The values $c$ for which $P_c$ has an attracting fixed point form a cardioid $C \subset M$, and $\partial C \subset \partial M$. If $c \in C$, then $\mathcal{J}_c$ is a quasicircle.*

*Proof.* The fixed points of $P_c$ are at $z_c = (1 \pm \sqrt{1-4c})/2$, and the multiplier at $z_c$ is $\lambda(c) = 2z_c$. Since $c = \lambda/2 - \lambda^2/4$, the condition $|\lambda(c)| < 1$ corresponds in the $c$-plane to the cardioid

$$C = \{\lambda/2 - \lambda^2/4 : |\lambda| < 1\}.$$

It is a subset of $M$, called the *main cardioid* of $M$. Since the function $\lambda/2 - \lambda^2/4$ is one-to-one on the closed unit disk, we have the uniqueness assertion of the theorem. According to the discussion at the end of Section VI.2, the Julia set $\mathcal{J}_c$ is a quasicircle for $c \in C$.

Let $W$ be the component of the interior of $M$ containing $C$. By Theorem 1.2, the polynomials $f_n(c) = P_c^n(0)$ are uniformly bounded on $W$. They converge to the attracting fixed point $z_c$ on $C$, hence on account of analyticity to the fixed point $z_c$ on all of $W$. However, if $c \notin \overline{C}$, then $z_c$ is repelling, and we cannot have $P_c^n(0) \to z_c$ unless $P_c^n(0) = z_c$ for $n$ large. Since this can occur only on a countable set, we conclude that $W = C$. $\square$

The figure in Section VI.2 corresponds to $c = i/2$, which is inside the main cardioid. Another example is given in Figure 6 to follow.

The multiplier function $\lambda(c) = 1 - \sqrt{1-4c}$ maps the closure of the cardioid homeomorphically onto the closed unit disk. The boundary $\partial C$ consists of precisely the parameter values $c$ for which $P_c$ has a neutral fixed point. These are favorite choices for generating computer pictures. The Siegel disks appearing in Section V.1 correspond

FIGURE 2. Attracting flowers.

to a value $c$ on the boundary of the main cardioid with multiplier $\lambda = e^{2\pi i \theta}$, where $\theta = (\sqrt{5} - 1)/2$ is the golden mean, as discussed in Section V.1. For a multicolored picture of the Siegel disk corresponding to this $c$-value, see Map 25 on page 77 of [**PeR**]. The filled-in Julia sets corresponding to parabolic fixed points with multipliers $\lambda = \exp(2\pi i/3)$ and $\lambda = \exp(2\pi i/5)$ are depicted in Figure 2.

Now we consider the more general case of attracting cycles.

THEOREM 1.4. *Suppose there is an attracting cycle of length $m$ for $P_a$. Then $a$ belongs to the interior of $\mathcal{M}$. If $W$ is the component of the interior of $\mathcal{M}$ containing $a$, then $P_c$ has an attracting cycle $\{z_1(c), ..., z_m(c)\}$ of length $m$ for all $c \in W$, where each $z_j(c)$ depends analytically on $c$.*

*Proof.* Let $z_1(a)$ be an attracting periodic point of period $m$ for $a$. Applying the implicit function theorem to $Q(z, c) = P_c^m(z) - z$, we obtain an attracting periodic point $z_1(c)$ for $P_c$ of period $m$, which depends analytically on $c$ in a neighborhood of $a$. In particular, $a$ belongs to the interior of $\mathcal{M}$, say to the component $W$. The sequence $f_j(c) = P_c^{jm}(0)$ is bounded hence normal on $W$, and it converges at $a$ to some point in the cycle of $z_1(a)$, say to $z_1(a)$ itself. Since $z_1(c)$ is attracting, $f_j(c)$ converges to $z_1(c)$ for $c$ near $a$. By normality $f_j(c)$ converges on $W$ to some analytic function $g(c)$, which satisfies

$Q(g(c), c) = 0$ near $a$ hence on $W$. Now $g(c)$ can be a repelling periodic point for fixed $c \in W$ only if $P_c^{jm}(0)$ actually coincides with $g(c)$ for $j$ large. This occurs on an at most countable set. Since the multiplier $\lambda(c)$ of the cycle is analytic, we conclude that $|\lambda| < 1$ on $W$, that is, the cycle of $g(c)$ is attracting for all $c \in W$. Further, the period of $g(c)$ is exactly $m$ for all $c \in W$ (as in Lemma III.2.5). The functions $P_c^i(g(c))$, $0 \le i < m$, give analytic selections of the points of the cycle. $\square$

The set of $c$ for which $P_c$ has an attracting cycle was studied by R. Brooks and J. Matelski in 1978. The components of the interior of $\mathcal{M}$ associated with attracting cycles are called the *hyperbolic components* of the interior of $\mathcal{M}$ since, by Theorem V.2.2, $P_c$ is hyperbolic precisely when $c \notin \mathcal{M}$ or $P_c$ has an attracting cycle. It is not known whether the hyperbolic components fill out the interior of $\mathcal{M}$, but we will soon see that they are dense in $\mathcal{M}$. Brooks and Matelski arrived at the work of Fatou and Julia from a problem on discrete subgroups of $PSL(2, \mathbf{C})$, and their rough computer picture was the first picture of the Mandelbrot set. Working about the same time and completely independently, B. Mandelbrot obtained progressively sharper pictures of the Mandelbrot set, which called attention to the intricate nature of the set.

One of the hyperbolic components is the main cardioid $C$, corresponding to cycles of length one. For cycles of length two, we have

$$Q(z, c) = (z^2 + c)^2 + c - z = (z^2 - z + c)(z^2 + z + c + 1).$$

Discarding fixed points, we obtain the equation $z^2 + z + c + 1 = 0$ for periodic points of period two. The multiplier is given by

$$(P_c^2)'(z) = 4z^3 + 4zc \equiv 4c + 4 \quad \text{modulo } (z^2 + z + c + 1).$$

Thus there is an attracting two-cycle if and only if $|4c + 4| < 1$, and we obtain a single hyperbolic component which is a disk centered at $-1$ of radius $1/4$, tangent to the main cardioid. Note again that the multiplier $\lambda(c)$ maps the hyperbolic component conformally onto the open unit disk. We show in Section 2 that each hyperbolic component of the interior of $\mathcal{M}$ contains a unique point $c$ such that $P_c$ has a superattracting cycle, and that the corresponding multiplier map $c \to \lambda(c)$ maps the component conformally onto the unit disk.

Of special interest are the parameter values $c$ for which $P_c$ has a superattracting cycle. The critical point 0 must belong to the cycle,

FIGURE 3. Airplane set, $c \approx -1.754878$.

so these are precisely the $c$'s for which 0 is periodic, that is, the solutions of $P_c^n(0) = 0$. The first few polynomials are

$$\begin{aligned}
P_c^1(0) &= c, \\
P_c^2(0) &= c + c^2, \\
P_c^3(0) &= c + c^2 + 2c^3 + c^4, \\
P_c^4(0) &= c + c^2 + 2c^3 + 5c^4 + 6c^5 + 6c^6 + 4c^7 + c^8.
\end{aligned}$$

For $n = 1$, the equation reduces to $c = 0$, and we obtain the super-attracting fixed point at 0 for $P_0(z) = z^2$. For $n = 2$, the equation becomes $c^2 + c = 0$. This gives an additional solution $c = -1$, for which $P_{-1}(z) = z^2 - 1$ has a superattracting cycle $0 \to -1 \to 0$ of period two. See Figure 8.

The superattracting cycles of period three correspond to the solutions of $P_c^3(0) = 0$, excluding $c = 0$. These are the roots of $c^3 + 2c^2 + c + 1 = 0$, which are approximately $-1.755$ and $-0.123 \pm 0.749i$. The real root belongs to the largest component of the interior of $\mathcal{M}$ located on the "main antenna" towards the left of $\mathcal{M}$. The filled-in Julia set for the real root is the "airplane set" pictured in Figure 3. The two complex roots belong to the largest components of the interior of $\mathcal{M}$ above and below the main cardioid. The filled-in Julia set corresponding to a complex root is pictured in Figure 4. This set is known as "Douady's rabbit."

THEOREM 1.5. *The values of $c \in \mathcal{M}$ corresponding to superattracting cycles cluster on the entire boundary $\partial \mathcal{M}$. In particular, the interior of $\mathcal{M}$ is dense in $\mathcal{M}$.*

*Proof.* Let $U$ be a disk that meets $\partial \mathcal{M}$, such that $0 \notin U$. Suppose $U$ contains no $c$ for which 0 is periodic. Consider a branch of $\sqrt{-c}$ defined on $U$. We have $P_c^n(0) \neq \sqrt{-c}$, or else $P_c^{n+1}(0) = 0$ and 0 is periodic. Thus $f_n(c) = P_c^n(0)/\sqrt{-c}$ omits the values $0, 1, \infty$ on

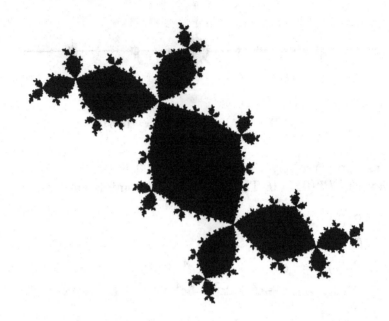

FIGURE 4. Douady's rabbit, $c \approx -0.122561 + 0.744862i$.

$U$, hence is a normal sequence on $U$. But since $U$ meets $\partial \mathcal{M}$, it contains both points $c$ with $f_n(c)$ bounded and with $f_n(c) \to \infty$, so the sequence cannot be normal. $\square$

Let us see how the Julia set changes shape as $c$ moves along the real axis. If we move $c$ to the right of $1/4$, it leaves the Mandelbrot set and the Julia set becomes totally disconnected. At the right of Figure 5 we have a picture of $\mathcal{J}_c$ for $c = 0.251$. The black points are those for which $|P_c^n(0)| \leq 4$ for $n \leq 500$. At the left of Figure 5 is the cauliflower set, corresponding to $c = 1/4$. In this case $\mathcal{J}$ is a simple closed Jordan curve (Section V.4), though it cannot be a quasicircle due to the cusps.

We show in Figure 6 the Julia set $\mathcal{J}_c$ for $c = -3/5$. We are back in the main cardioid, so $\mathcal{J}_c$ is a quasicircle, symmetric with respect to $\mathbf{R}$. Moving back to the left, at the left edge of the main cardioid we arrive at the point $c = -3/4$ with multiplier $-1$, so that $P_c$ has a parabolic fixed point $-1/2$. There are two petals at $-1/2$, which cycle back and forth. The filled-in Julia set is pictured in Figure 7.

When we continue to the left of $-3/4$, the fixed point bifurcates to an attracting cycle of length two, corresponding to the two petals.

FIGURE 5. $c = 0.25$ (cauliflower set), and $c = 0.251$ (totally disconnected).

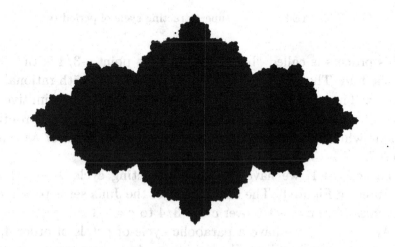

FIGURE 6. $c = -0.6$, quasicircle.

FIGURE 7. $c = -0.75$, parabolic cycle of period two.

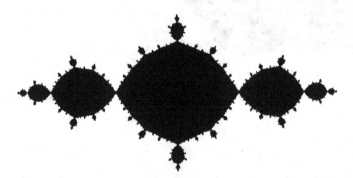

FIGURE 8. $c = -1$, superattracting cycle of period two.

This process is called "budding," and the point $-3/4$ is the "root" of the bud. There is budding at each point of $\partial C$ with rational multiplier. The parabolic fixed point with multiplier a primitive $m$th root of unity splits into an attracting cycle of length $m$, sprouting a "bud" which is a hyperbolic component of the interior of $\mathcal{M}$ tangent to $\partial C$.

For $c = -1$, we have the superattracting cycle $0 \to -1 \to 0$ pictured in Figure 8. The basic shape of the Julia set is preserved as we cross from $c = -3/5$ over $c = -3/4$ to $c = -1$.

At $c = -5/4$ we have a parabolic cycle of petals of order 4, and there is further budding. Continuing to decrease $c$ gives a sequence $c_0 > c_1 > c_2 > \cdots$ of parameter values corresponding to parabolic cycles of order $2^n$. In the complementary intervals, $P_c$ has attracting cycles of order $2^n$. This behavior is known as the *period doubling* of Feigenbaum, and $c_n \to c_\infty \approx -1.401$. In the interval $[-2, c_\infty]$, periods of many different orders occur.

A point $c \in \mathcal{M}$ is called a *Misiurewicz point* if 0 is strictly preperiodic, that is, $P_c^n(0) = P_c^k(0)$ for some $n > k > 0$, but $P_c^n(0) \neq 0$ for all $n \geq 1$. By Theorem V.2.3, these are exactly the points where $P_c$ is subhyperbolic without being hyperbolic on $\mathcal{J}_c$. By Theorem V.4.2, the Julia set $\mathcal{J}_c$ is a dendrite, and, by Theorem VII.3.1, $\overline{\mathbf{C}} \backslash \mathcal{J}_c$ is a John domain. Since $P_c$ is expanding with respect to some metric, there are no neutral cycles, and the cycle of $P_c^k(0)$ is repelling.

The Misiurewicz points are located in a complicated pattern in the antenna area of $\mathcal{M}$. Both $c = -2$ and $c = i$ are Misiurewicz points. We have seen that the Julia set of $z^2 - 2$ is $[-2, 2]$. The Julia set of $z^2 + i$ is pictured in Section V.4. Since the Misiurewicz points are the zeros of a sequence of polynomials, they are countable. The proof of Theorem 1.5 can be modified to show that they accumulate on all of $\partial \mathcal{M}$. We will see this in another way in Sections 5 and 6, where we show that these points belong to $\partial \mathcal{M}$ and are terminal points of arcs in the complement of $\mathcal{M}$.

## 2. The Hyperbolic Components of $\mathcal{M}$

We return to the polynomial $Q(z, c) = P_c^m(z) - z$ in $z$ and $c$. The zero set $V$ of $Q$ is a one-dimensional complex analytic variety in $\mathbf{C}^2$. It consists of all $(z, c)$ such that $z$ is a periodic point of $P_c$ whose period divides $m$. Note that $Q$ is a monic polynomial in both $z$ and in $c$, so that if $(z, c) \in V$ and one of $|z|$ or $|c|$ tends to infinity, then so does the other. The projection of $V$ onto the $c$-plane is a branched covering map with $2^m$ sheets.

Fix one of the hyperbolic components $W$ of the interior of $\mathcal{M}$ that corresponds to attracting cycles of length $m$, and let $z_1(c)$ be the analytically varying periodic point of $P_c$ from Theorem 1.4. Let $\tilde{W}$ be the subset of $V$ of points $(z_1(c), c)$ for $c \in W$, and let $V_0$ be the irreducible branch of $V$ containing $\tilde{W}$. We can view $V_0$ as a Riemann surface spread over the $c$-plane, on which $(P_c^m)'(z)$ is analytic. Let $\tilde{V}$ be the subset of $V_0$ where $|(P_c^m)'(z)| < 1$. Then $\partial \tilde{V}$ consists of piecewise analytic curves on which $|(P_c^m)'(z)| = 1$. As $c \in W$ tends to $a \in \partial W$, we have $(z_1(c), c) \to \partial \tilde{V}$, or else we could continue $z_1(c)$ analytically to a neighborhood of $a$ to give attracting periodic points, contradicting $a \in \partial \mathcal{M}$. Thus $\tilde{W}$ is a connected component of

$\tilde{V}$. The projections of the analytic arcs in $\partial \tilde{W}$ are piecewise analytic. (At worst they might develop cusps where $(P_c^m)'(z) = 1$.) Thus $\partial W$ is piecewise analytic. Since $W \cup \partial W$ has connected complement, $\partial W$ has no multiple points, and $\partial W$ must be a simple closed Jordan curve. Thus $\partial \tilde{W}$ is also a simple closed Jordan curve, and the branch $z_1(c)$ extends continuously from $W$ to $W \cup \partial W$ (including to the cusps).

Let $\rho(c)$ denote the multiplier of the attracting cycle for $c \in W$. Since $\rho(c) = (P_c^m)'(z_1(c))$, $\rho(c)$ extends continuously to $\partial W$, where it has unit modulus. Consequently, the extended multiplier function $\rho(c)$ is a continuous finite-to-one map of $\overline{W}$ onto the closed unit disk $\overline{\Delta}$, which maps $W$ analytically onto $\Delta$. Our main goal in this section is to prove the theorem of Douady–Hubbard–Sullivan (see [Do1]) that this covering is simple, so that the multiplier function $\rho(c)$ parametrizes $\overline{W}$.

THEOREM 2.1. *If $W$ is a hyperbolic component of the interior of $\mathcal{M}$, then the multiplier $\rho(c)$ of the attracting cycle of $P_c$, $c \in W$, maps $W$ conformally onto the open unit disk $\Delta$. It extends continuously to $\partial W$ and maps $\overline{W}$ homeomorphically onto the closed disk $\overline{\Delta}$.*

*Proof.* Fix $a \in W$. Let $U_1, ..., U_{m-1}, U_m = U_0$ be the cycle of components of $\mathcal{F}$ containing the attracting cycle. We assume $U_0$ contains the critical point 0, so $P_a$ is two-to-one from $U_0$ onto $U_1$, and $P_a$ maps each other $U_j$ conformally onto $U_{j+1}$. Let $\varphi$ map $U_0$ conformally to $\Delta$, with the fixed point of $P_a^m$ going to 0. Then $\varphi \circ P_a^m \circ \varphi^{-1}$ is a Blaschke product of degree two, mapping 0 to 0. After normalization of $\varphi$, the Blaschke product has the form

$$(\varphi \circ P_a^m \circ \varphi^{-1})(\zeta) = B_\lambda(\zeta) = \zeta \frac{\zeta + \lambda}{1 + \overline{\lambda}\zeta}, \qquad |\zeta| < 1,$$

for some $\lambda \in \Delta$. Since $\lambda$ is the multiplier of the fixed point at 0 for $B_\lambda(\zeta)$, and this is invariant, we have $\lambda = \rho(a)$.

The idea now is to reverse this procedure. For $|\lambda| < 1 - \varepsilon$, we combine the dynamics of $P_a(z)$ outside the $U_j$'s with $B_\lambda$ inside, to obtain $c(\lambda)$ so that $P_{c(\lambda)}$ has multiplier $\rho(c(\lambda)) = \lambda$, and so that $c(\lambda)$ depends continuously (!) on $\lambda$. Since $c(\lambda)$ must have a branch point at a critical value of $\rho$, the continuity of $c(\lambda)$ shows that $\rho(c)$ has no critical points, and hence $\rho$ is one-to-one. The construction of $c(\lambda)$ is done by quasiconformal surgery as in the proof of Theorem VII.4.1, except that here the Beltrami equation is solved with a continuous parameter $\lambda$. Attention must also be paid to the complication that there is an invariant cycle rather than a fixed component.

So fix $\varepsilon > 0$ small, let $1 - \varepsilon < r < 1$, and denote $\Delta_r = \{|\zeta| < r\}$. We assume that $r$ is so close to 1 that for each $\lambda$, $|\lambda| < 1 - \varepsilon$, the curve $B_\lambda^{-1}(\partial \Delta_r)$ is a simple closed Jordan curve in the annulus $r < |\zeta| < 1$ mapped two-to-one by $B_\lambda$ onto the circle $\partial \Delta_r$. Set $A_0 = U_0 \cap P_a^{-m}(\varphi^{-1}(\Delta_r))$. Then $\varphi^{-1}(\Delta_r) \subset A_0$, and $P_a^m$ is a two-to-one cover of $A_0$ over $\varphi^{-1}(\Delta_r)$. Define $\varphi_\lambda : \partial A_0 \to B_\lambda^{-1}(\partial \Delta_r)$ so that $B_\lambda \circ \varphi_\lambda = \varphi \circ P_a^m$, choosing the branch so that $\varphi_\lambda$ is continuous in $\lambda$. Extend $\varphi_\lambda$ smoothly to a diffeomorphism from $A_0$ to $B_\lambda^{-1}(\partial \Delta_r)$, to vary smoothly also with $\lambda$, so that $\varphi_\lambda = \varphi$ on $\varphi^{-1}(\Delta_r)$. Define $A_j \subset U_j$ for $1 \le j \le m$ so that $A_m = A_0$ and $P_a$ maps each $A_j$ conformally onto $A_{j+1}$, $1 \le j < m$. Define $g_\lambda(z) = P_a(z)$ for $z \notin \cup A_j$, and define $g_\lambda(z)$ on $\cup A_j$ so that it maps each $A_j$ into $A_{j+1}$, and so that $g_\lambda^m(z) = \varphi^{-1}(B_\lambda(\varphi_\lambda(z)))$ on $A_0$. Thus $g_\lambda(z) = P_a(z)$ on $A_1 \cup \cdots \cup A_{m-1}$, and $g_\lambda(z) = P_a^{-m+1}(g_\lambda^m(z))$ on $A_0$. Consider the ellipse field that is circles outside of $\cup A_j$ and that is invariant under $g_\lambda$. Under iteration by $g_\lambda$, points hit the annular region where $g_\lambda$ is not conformal at most once, so the ellipse field is distorted at most once on any orbit, and consequently the Beltrami coefficients $\mu_\lambda$ are bounded strictly less than 1 in modulus. Moreover $\mu_\lambda$ moves continuously with $\lambda$. Let $\psi_\lambda$ solve the Beltrami equation, with the usual normalization $\psi_\lambda(z) = z + o(1)$ at $\infty$. Then $f_\lambda = \psi_\lambda \circ g_\lambda \circ \psi_\lambda^{-1}$ is analytic, and $f_\lambda(z) = z^2 + \mathcal{O}(1)$ at $\infty$, so $f_\lambda(z) = z^2 + c(\lambda) = P_{c(\lambda)}(z)$. Since $g_\lambda$ has an attracting cycle of length $m$, so does $f_\lambda$, and the multipliers are the same. Thus $\rho(c(\lambda)) = \lambda$. Finally the solutions $\psi_\lambda$ depend continuously on the parameter $\lambda$, so that $c(\lambda)$ depends continuously on $\lambda$. By the initial remarks we have proved the theorem. $\square$

If $P_a$ has a neutral cycle of length $m$, then $a \in \partial W$ for some hyperbolic component $W$ associated with attracting cycles of length $m$. Indeed, let $z_0$ be a point in the cycle, and let $V_0$ be an irreducible branch of $V$ containing $(z_0, a)$. Then $(P_c^m)'(z)$ cannot be constant on $V_0$, or else since $V_0$ projects onto $\mathbf{C}$ there would be a neutral cycle for all $P_c$, an absurdity. Thus $(z_0, a)$ is in the boundary of some component $\tilde{W}$ of $\tilde{V}$ (defined earlier). The projection $W$ of $\tilde{W}$ into the $c$-plane is a hyperbolic component of the interior of $\mathcal{M}$, and $a \in \partial W$.

If $W$ is associated with attracting cycles of length $m$, then every

$c \in \partial W$ with $\rho(c) \neq 1$ has a neutral cycle of length $m$, which is the analytic continuation of the attracting cycles of $W$. If $\rho(c) = 1$, then $P_c$ has a parabolic cycle whose length divides $m$. If the length of the cycle of $P_c$ is a proper divisor of $m$, then $c$ lies on the boundary of another hyperbolic component, from which $W$ is obtained by "budding." The point on $\partial W$ where $\rho(c) = 1$ is called the *root* of $W$, and the point of $W$ where $\rho(c) = 0$ is called the *center* of $W$.

EXAMPLE. There are three superattracting cycles of period three, hence three hyperbolic components corresponding to attracting three-cycles. One of these is the largest bulb on the top of the main cardioid. At the root of the component, the cycle of period three coalesces to a parabolic fixed point with multiplier $e^{2\pi i/3}$, depicted in Figure 2 of Section 1. The second hyperbolic component is located in the symmetric position below the main cardioid. The third hyperbolic component is the largest component centered on the real axis between $-2$ and the Feigenbaum point. For a color close-up of this component of the Mandelbrot set, see Map 32 on p. 89 of [**PeR**]. The root of the component is a cusp, and it corresponds to a parabolic cycle of length three and multiplier 1.

## 3.   Green's Function of $\mathcal{J}_c$

For the analytic study of $P_c$, the conjugation map $\varphi_c(z)$ at $\infty$ is the natural tool,
$$\varphi_c(P_c(z)) = \varphi_c(z)^2.$$
It is uniquely determined and has the form $\varphi_c(z) = z + o(1)$ at $\infty$. Recall from the superattracting case of Chapter II that
$$\varphi_c(z) = \lim_{n \to \infty} P_c^n(z)^{2^{-n}},$$
which can be written
$$\varphi_c(z) = z \prod_{n=1}^{\infty} \left( \frac{P_c^n(z)}{(P_c^{n-1}(z))^2} \right)^{2^{-n}} = z \prod_{n=0}^{\infty} \left( 1 + \frac{c}{(P_c^n(z))^2} \right)^{2^{-n-1}}.$$

The functional equation for $\varphi_c$ allows us to extend $\log |\varphi_c(z)|$ harmonically to $A_c(\infty)$, where it coincides with Green's function $G_c(z)$ for $A_c(\infty)$ with pole at $\infty$. From the discussion in Section II.4 we have the following.

THEOREM 3.1. *The logarithmic capacity of the Julia set $\mathcal{J}_c$ is equal to 1 for all $c$.*

In this case the functional equation for Green's function is

$$G_c(P_c(z)) = 2G_c(z), \qquad z \in A_c(\infty).$$

By the discussion in Section III.4, there are two cases that occur. If $c \in \mathcal{M}$, there are no critical points of $P_c$ (or $G_c$) in $A_c(\infty)$, and the equation $\varphi_c(z) = \sqrt{\varphi_c(P_c(z))}$ allows us to extend $\varphi_c(z)$ to a conformal map of all of $A_c(\infty)$ onto $\{|\zeta| > 1\}$. On the other hand, if $c \notin \mathcal{M}$, then $0 \in A_c(\infty)$, and $\varphi_c(z)$ extends analytically to the exterior $\{G_c > G_c(0)\}$ of the level curve of Green's function passing through the critical point 0, mapping it conformally onto $\{|\zeta| > e^{G_c(0)}\}$. The level curve forms a figure-eight with cusp at 0, as in Section III.4, and $\varphi_c(z)$ approaches different values as $z$ approaches 0 from different sides.

Recall that Green's lines are the orthogonal lines to the level curves of Green's function. If $c \in \mathcal{M}$, these are just the curves in $A_c(\infty)$ mapped by $\varphi_c(z)$ to rays in conjugation space. Even if $c \notin \mathcal{M}$, they are well-defined, and they cluster at the boundary of $A_c(\infty)$, except for at most countably many that meet a critical point of $G_c$. We define $\mathcal{R}(\theta, \mathcal{K}_c)$, called an *external ray of $\mathcal{K}_c$*, to be the Green's line that corresponds to the ray $Re^{2\pi i\theta}$ in conjugation space, that is, the Green's line with initial segment $\varphi_c^{-1}(Re^{2\pi i\theta})$, $e^{G_c(0)} < R < \infty$. Thus Green's lines are parametrized by a parameter $\theta$ from 0 to 1 (mod 1), where as usual we measure angles in turns rather than radians. The action of $P_c$ on Green's lines corresponds to doubling the angle in conjugation space, always reduced mod 1,

$$z \in \mathcal{R}(\theta, \mathcal{K}_c) \quad \Longrightarrow \quad P_c(z) \in \mathcal{R}(2\theta, \mathcal{K}_c).$$

The dynamics of $P_c$ on dynamic space are closely related to the dynamics of the doubling transformation in $\theta$-space.

If the ray $\mathcal{R}(\theta, \mathcal{K}_c)$ terminates at a point $z_0 \in \mathcal{K}_c$, we say that $\theta$ is an *external angle of $\mathcal{K}_c$ at $z_0$*. If $c \in \mathcal{M}$ and $P_c$ has an attracting or parabolic cycle, then all external rays terminate, and all points of $\mathcal{J}_c$ have at least one external angle. If $c \notin \mathcal{M}$, then $\mathcal{K}_c = \mathcal{J}_c$ is totally disconnected, so $\mathcal{R}(\theta, \mathcal{J}_c)$ terminates at a point of $\mathcal{J}_c$ as soon as it is defined.

We extend the definition of Green's function $G$ for a domain $D$ by declaring it to be zero in the complement of $D$. With this convention in mind, we show that $G_c$ is Hölder continuous.

THEOREM 3.2. *For every $A > 0$, there exists $\alpha = \alpha(A) > 0$ so that Green's function $G_c$ is uniformly $\alpha$-Hölder continuous for $|c| \leq A$.*

*Proof.* Assume $A \geq 10$. Let $z \in \mathbf{C} \backslash \mathcal{K}_c$ and let $\delta(z) = \operatorname{dist}(z, \mathcal{K}_c)$. Let $z_0 \in \mathcal{K}_c$ be the point closest to $z$, and let $S$ be the straight line segment from $z_0$ to $z$. Take $N = N(z)$ to satisfy $|P^n(w)| < A$ for all $w \in S$ and all $n < N$, while $|P^N(z_1)| \geq A$ for some $z_1 \in S$. Using $P'(z) = 2z$ and the chain rule, we see that $|(P^n)'(w)| \leq (2A)^n$ for all $n < N$ and $w \in S$. Also, $|P^n(z_0)| < 2\sqrt{A}$ for all $n$, or else the iterates would escape to $\infty$. The mean value theorem then implies

$$|P^N(z_1)| \leq 2\sqrt{A} + 2^N A^N \delta(z_1).$$

But $|P^N(z_1)| \geq A$, hence $2^N A^N \delta(z_1) \geq 1$, and $2^N A^N \delta(z) \geq 1$. For $\alpha = \log 2/(\log 2 + \log A)$ we then have $\delta(z)^\alpha > 2^{-N}$, so that

$$G(z) = G(P^N(z))2^{-N} \leq M\delta(z)^\alpha,$$

where $M$ depends only on $A$.

Consider two points $z_1, z_2$ and suppose $\delta(z_1) \geq \delta(z_2)$. We want to prove

$$|G(z_1) - G(z_2)| \leq C|z_1 - z_2|^\alpha.$$

If $|z_1 - z_2| > \frac{1}{2}\delta(z_1)$ this follows from the estimate above. If $|z_1 - z_2| \leq \frac{1}{2}\delta(z_1)$, just use the fact that $G(z)$ is a positive harmonic function in the disk $\Delta(z_1, \delta(z_1))$, to conclude

$$|G(z_1) - G(z_2)| \ \leq \ C_0 M \delta(z_1)^\alpha |z_1 - z_2|/\delta(z_1) \ \leq \ C|z_1 - z_2|^\alpha. \quad \square$$

The preceding proof extends easily to polynomials of higher degree, and the corresponding theorem is due to N. Sibony (seminar talk, 1981). The idea of the proof has been used by J.E. Fornaess and N. Sibony (1992) to obtain Hölder continuity of Green's function associated with certain polynomial mappings (complex Hénon mappings) on $\mathbf{C}^2$. It can be shown that Green's function of a domain with uniformly perfect boundary is Hölder continuous, so Theorem III.3.3 leads to Hölder continuity also in the case of iteration of rational functions.

The Julia sets $\mathcal{J}_c$ for $|c| < A$ have Hausdorff dimension at least as large as the exponent $\alpha(A)$ of the preceding theorem. This follows from a theorem of Carleson (1963), that a compact subset of $\mathbf{R}^m$ is a set of removable singularities for $\alpha$-Hölder continuous harmonic functions if and only if it has zero $(m-2+\alpha)$-dimensional Hausdorff measure.

If $c \in \mathcal{M}$, we can prove directly that $G_c(z)$ is Hölder continuous with exponent $\alpha = 1/2$. In this case the map $f(w) = 1/\varphi_c^{-1}(1/w)$ is a normalized univalent function on $\Delta$, belonging to $\mathcal{S}$. The estimate of Theorem I.1.7 for $f$ leads immediately to an estimate of the form $G_c(z) \leq C\delta(z)^{1/2}$ for $z$ in a neighborhood of $\mathcal{K}_c$, and the Hölder estimate follows as in the preceding proof. Note that the exponent $1/2$ is sharp, as can be seen by taking $c = -2 \in \mathcal{M}$, for which $\mathcal{J}_c$ is the interval $[-2, 2]$.

**THEOREM 3.3.** *If $c_n \to c$, then the corresponding Green's functions $G_{c_n}(z)$ converge uniformly on $\mathbf{C}$ to $G_c(z)$. Thus $G_c(z)$ is jointly continuous in $c$ and $z$.*

*Proof.* The uniform Hölder estimates show that the sequence of functions $G_{c_n}(z)$ is equicontinuous on compact sets. Let $H$ be a uniform limit on compacta of a subsequence. Then $H$ is continuous, and $H$ is harmonic on the set $\{H > 0\}$. By the maximum principle, there are no bounded components of the set $\{H > 0\}$, just one unbounded component. Since the conjugating functions $\varphi_c$ depend analytically on $c$, $G_{c_n}(z) = \log|\varphi_{c_n}(z)|$ converges uniformly to $G_c(z) = \log|\varphi_c(z)|$ for $|z|$ large. It follows that $H = G_c$ on $A_c(\infty)$, hence everywhere. Since the limit $H$ is unique, we have uniform convergence of the original sequence. $\square$

## 4.  Green's Function of $\mathcal{M}$

Suppose $c \in \mathbf{C} \backslash \mathcal{M}$. As long as $G_c(z) > G_c(0) > 0$, the function $\varphi_c(z)$ is well-defined and analytic. This holds in particular for $z = c$, since $G_c(c) = 2G_c(0) > G_c(0) > 0$. Hence

$$\Phi(c) = \varphi_c(c) = c \prod_{n=0}^{\infty} \left(1 + \frac{c}{P_c^n(c)^2}\right)^{2^{-n-1}}$$

is analytic, $\Phi$ has a simple pole at $\infty$, and $\log|\Phi(c)| = G_c(c) = 2G_c(0)$. By Theorem 3.3, $G_c(c) \to 0$ as $c \to M$, and consequently $|\Phi(c)| \to 1$ as $c \to M$. By the argument principle, $\Phi$ assumes every value in $\mathbf{C}\backslash\overline{\Delta}$ exactly once on $\mathbf{C}\backslash M$, and $\Phi$ maps $\overline{\mathbf{C}}\backslash M$ conformally onto the exterior of the closed unit disk $\overline{\mathbf{C}}\backslash\overline{\Delta}$. In particular, $\overline{\mathbf{C}}\backslash M$ is simply connected, and we obtain the theorem of Douady and Hubbard (1982), proved independently by N. Sibony (cf. [**DH1**]), that $M$ is connected.

Since $\Phi(c) = c + \mathcal{O}(1)$ as $c \to \infty$, Green's function of $\overline{\mathbf{C}}\backslash M$, which is $\log|\Phi(c)|$, has the form $\log|c| + o(1)$ at $\infty$. Hence Robin's constant is 0, and $M$ has capacity 1. We state these results formally.

THEOREM 4.1. *The Mandelbrot set M is connected and has logarithmic capacity equal to 1.*

In analogy to the external rays of $\mathcal{K}_c$, we define the *external rays of M* to be Green's lines for $\overline{\mathbf{C}}\backslash M$, that is, the inverse images of radial lines under $\Phi$. These are denoted by $\mathcal{R}(\theta, M)$, $\theta$ between 0 and 1 (mod 1). If the ray terminates at $c \in M$, we say that $\theta$ is an *external angle of M* at $c$. Observe that the external rays of $M$ are related to the external rays of $\mathcal{K}_c$ in the sense that

$$c \in \mathcal{R}(\theta, M) \iff c \in \mathcal{R}(\theta, \mathcal{K}_c).$$

It is not known whether every external ray of $M$ terminates, nor is it known whether the inverse $\psi$ of $\Phi$ extends continuously to map $\partial\Delta$ onto $\partial M$. This latter question is equivalent by Carathéodory's theorem to $M$ being locally connected. A great deal of work has gone into this question. It may be of interest to note that this would follow if there were some uniform estimate of escape times for $c$ in terms of its distance to $M$.

THEOREM 4.2. *Let $\delta(c) = \mathrm{dist}(c, M)$. If there exists a decreasing function $F(x) \geq 1$ on $[0, 1]$ such that*

$$\int_0^1 F(x)dx < \infty$$

*and*

$$|P_c^N(c)| > 5 \quad \text{for} \quad N \geq F(\delta(c)),$$

*then $M$ is locally connected.*

*Proof.* Let $\psi : \{|\zeta| > 1\} \to \mathbf{C} \backslash \mathcal{M}$ be the inverse map to $\Phi(c)$. If we can show that $|\psi'(\zeta)| \le H(|\zeta|)$ for some decreasing function $H(r)$ on $(1, \infty)$ such that

$$\int_1^2 H(r) dr < \infty,$$

then $\psi$ is continuous up to the boundary, and we are done.

Let $N$ be the first integer larger than $F(\delta(c))$. Since $G(P_c(z)) = 2G(z)$, $|P_c^N(c)| > 5$ implies that $G_c(c) > C_0 2^{-N}$. For $\zeta = \Phi(c)$ we have $G_c(c) = \log |\zeta| \sim |\zeta| - 1$, so

$$2^{-F(\delta(c))} \le 2^{-N+1} \le C_1(|\zeta| - 1).$$

The Koebe one-quarter theorem gives $|\psi'(\zeta)|(|\zeta| - 1) \le 4\delta(c)$, so that

$$2^{-F(|\psi'(\zeta)|(|\zeta| - 1)/4)} \le C_1(|\zeta| - 1).$$

Setting $t = C_1(|\zeta| - 1)$ and solving $s = 2^{-F(st)}$ for $s$, we find that

$$\frac{1}{4C_1}|\psi'(\zeta)| \le s \le \frac{1}{t} F^{-1}\left(\frac{1}{\log 2} \log \frac{1}{t}\right).$$

Thus the function

$$H(r) = \frac{4}{r-1} F^{-1}\left(\frac{1}{\log 2} \log \frac{1}{C_1(r-1)}\right)$$

has the desired properties, since

$$\int_1^2 H(r) dr = C_2 \int_A^\infty F^{-1}(x) dx \le C_2 \int_0^1 F(x) dx < \infty. \quad \square$$

For a hyperbolic situation, $N \sim C \log(1/\delta(c))$, which is much faster than we need. Even an escape time estimate $N \sim \delta(c)^{-1+\varepsilon}$ would be sufficient to prove $\mathcal{M}$ is locally connected.

It would be of interest to run computer experiments to estimate the escape time $N$ in terms of $\delta(c)$ (recommendation: start near the Feigenbaum point). Towards this end we mention an effective algorithm, whose idea is due to J. Milnor and W. Thurston (in [**Mi2**]; for subsequent developments see [**Pe**], [**Fi**]). The algorithm estimates $\delta(c)$ for $c \notin \mathcal{M}$. It is based on the approximation arising from the Koebe one-quarter theorem,

$$\delta(c) \sim \frac{|\Phi(c)| - 1}{|\Phi'(c)|},$$

as in Theorem I.1.4. Set $z_N(c) = P_c^N(0)$. We obtain by differentiation of $\log \Phi(c) = 2^{-N+1} \log \varphi_c(z_N(c))$ that

$$\frac{|\Phi'(c)|}{|\Phi(c)|} = 2^{-N+1} \frac{|\varphi_c'(z_N(c))|}{|\varphi_c(z_N(c))|} |z_N'(c)|.$$

The approximation $|\Phi(c)| - 1 \sim \log|\Phi(c)| = 2^{-N+1} \log|\varphi_c(z_N(c))|$ for $N$ large yields

$$\frac{|\Phi(c)| - 1}{|\Phi'(c)|} \sim \frac{|\varphi_c(z_N(c))| \log|\varphi_c(z_N(c))|}{|\Phi(c)||\varphi_c'(z_N(c))||z_N'(c)|}.$$

The approximations $|\Phi(c)| \sim 1$, $\varphi_c(z_N(c)) \sim z_N(c)$, $\varphi_c'(z_N(c)) \sim 1$ then lead to

$$\delta(c) \sim |z_N'(c)|^{-1}|z_N(c)| \log|z_N(c)|, \qquad c \notin M, \ c \text{ near } M, \ N \text{ large}.$$

The algorithm is then to compute successively $z_0 = 0$, $z_{j+1} = P_c(z_j)$, until $|z_N| > R$, and then compute the derivative of $z_N(c)$ by the chain rule, $z_0' = 0$, $z_{j+1}' = 2z_j z_j' + 1$. The approximator for $\delta(c)$ is $|z_N'|^{-1}|z_N| \log|z_N|$, and we can obtain good estimates for the error factor. The resolution of the pictures in [Pe] obtained by the distance-estimator algorithm is striking.

## 5.   External Rays with Rational Angles

The external rays with $\theta$ rational play a special role. We aim to show that these rays hit $M$ at well-defined points. If $\theta = p/q$ with $p$ and $q$ relatively prime, then the cases with $q$ odd or even are very different. If $q$ is odd, the ray terminates at a point $c \in M$ with the property that $0$ belongs to a parabolic component of $\mathcal{F}(P_c)$. If $q$ is even, it ends at a point $c$ for which $0$ is strictly preperiodic, that is, at a Misiurewicz point. The reason stems from a simple result on binary expansions.

Let $\theta \in [0, 1]$ have binary expansion $\theta = \sum_1^\infty \theta_j 2^{-j}$, where each $\theta_j$ is $0$ or $1$. The function $P_c$ acts on external rays by doubling $\theta$ (and reducing modulo 1). This corresponds to shifting the sequence of binary coefficients backwards (and chopping). Evidently the sequence $\{\theta_j\}$ is periodic if and only if $2^m \theta \equiv \theta \pmod 1$ for some $m \geq 1$. The sequence is preperiodic if and only if $2^n \theta \equiv 2^k \theta \pmod 1$ for some $n > k \geq 0$.

THEOREM 5.1. *Let $\theta = \sum_1^\infty \theta_j 2^{-j}$ be as above, with $0 < \theta \le 1$. Then $\theta$ is rational if and only if the sequence $\{\theta_j\}$ is preperiodic. If $\theta = p/q$, where $p$ and $q$ are relatively prime integers, then $q$ is odd if and only if $\{\theta_j\}$ is periodic, and $q$ is even if and only if $\{\theta_j\}$ is strictly preperiodic.*

*Proof.* Suppose $\theta = p/q = p/(2^k r)$ is rational, with $r$ odd, $p$ and $q$ relatively prime. Choose (by Euler's generalization of Fermat's little theorem) $m$ so that $2^m \equiv 1 \pmod r$. Then

$$2^{(m+k)}\theta = 2^k\theta + 2^k jr\theta = 2^k\theta + jp \equiv 2^k\theta,$$

so $\theta$ is preperiodic. If moreover $q$ is odd, then $k = 0$, and $\theta$ is periodic. Conversely, if $\theta$ is periodic, there are integers $m, n$ such that $2^m\theta = \theta + n$. Then $\theta = n/(2^m - 1)$ is rational, and the denominator must be odd. If $\theta$ is preperiodic, then $2^k\theta$ is periodic, and $\theta$ is again rational, with even denominator if $k \ge 1$. $\square$

THEOREM 5.2. *If $\theta$ is rational, then $\mathcal{R}(\theta, \mathcal{M})$ terminates at a point $a \in \partial\mathcal{M}$. If $\theta$ has even denominator, then $a$ is a Misiurewicz point, whereas if $\theta$ has odd denominator, then $P_a$ has a parabolic cycle.*

*Proof.* The proof, which is long, proceeds in outline as follows. If $\text{dist}(c, \mathcal{J}_c) \to 0$ as $c$ tends to $a$ along $\mathcal{R}(\theta, \mathcal{M})$, it is rather easy to see that $a$ is a Misiurewicz point and $\theta$ has even denominator. If $\text{dist}(c, \mathcal{J}_c)$ does not converge to $0$ as $c$ tends to $a$ along $\mathcal{R}(\theta, \mathcal{M})$, it is again easy to see that a parabolic component of the Fatou set is created for $P_a$. The main problem concerns the relation between the parabolic periodic points and the ray $\mathcal{R}(\theta, \mathcal{K}_a)$, which eventually yields the result that $\theta$ has odd denominator.

So let $\theta$ be rational with period $\ell$, say $2^{\ell+k}\theta \equiv 2^k\theta$, where $\ell \ge 1$ and $k \ge 0$ are minimal. If $c \in \mathcal{R}(\theta, \mathcal{M})$, then both $P_c^k(c)$ and $P_c^{\ell+k}(c)$ belong to $\mathcal{R}(2^k\theta, \mathcal{J}_c)$. Moreover, these points have uniformly bounded hyperbolic distance from each other in $\overline{\mathbf{C}} \setminus \mathcal{J}_c$. To see this, it suffices by Theorem I.4.2 to bound the hyperbolic distance between them in the simply connected subdomain $\{G_c > G_c(0)\}$ of $A_c(\infty)$. This subdomain is mapped conformally to the unit disk by $e^{G_c(0)}/\varphi_c(z)$, and the points are mapped to points with radii $r^{k+1}$ and $r^{\ell+k+1}$ on the same radial segment, where $r = e^{-G_c(0)}$. The hyperbolic distance between these points in the unit disk is bounded, independent of $r$ (for fixed $\ell$).

Fix a cluster point $a \in \partial \mathcal{M}$ of $\mathcal{R}(\theta, \mathcal{M})$. Suppose first there are points $c_j$ in $\mathcal{R}(\theta, \mathcal{M})$ converging to $a$ such that the distance from $c_j$ to $\mathcal{J}_{c_j}$ tends to 0. The lower estimate of Theorem I.4.3 featuring $\delta \log(1/\delta)$ shows that hyperbolic disks of uniformly bounded hyperbolic radius have euclidean diameters tending to 0 as the centers approach the boundary. Thus $|P_{c_j}^{\ell+k}(c_j) - P_{c_j}^k(c_j)| \to 0$, and in the limit we obtain $P_a^{\ell+k}(a) = P_a^k(a)$. Thus $a$ is preperiodic. Now the cycles containing 0 are superattracting and correspond to points in the interior of $\mathcal{M}$. Since $a \in \partial \mathcal{M}$ and since 0 is its only predecessor, $a$ must be strictly preperiodic, that is, $a$ is a Misiurewicz point. Thus $k \geq 1$, and $\theta$ has even denominator.

Next we analyze the case in which there are points $c_j$ in $\mathcal{R}(\theta, \mathcal{M})$ converging to $a$ such that the distances from $c_j$ to $\mathcal{J}_{c_j}$ are bounded away from 0. This situation occurs for instance for the rays $\mathcal{R}(0, \mathcal{J}_c)$ as $c > 1/4$ decreases to $1/4$, as illustrated in Figure 5, Section 1. The disconnected set $\mathcal{J}_c$ closes up, and the limit ray goes into the "interior." The point $c$ on the ray converges to $1/4$, strictly inside the filled-in Julia set $\mathcal{K}_{1/4}$, which has the parabolic fixed point $z = 1/2$ on its boundary.

Assuming $\text{dist}(c_j, \mathcal{J}_{c_j}) \geq \delta > 0$, Green's function $G_{c_j}(z)$ is positive and harmonic in $\Delta(c_j, \delta)$ and $G_{c_j}(c_j) \to 0$. Hence $G_a(z) = 0$ on $\Delta(a, \delta)$, and $a$ belongs to a bounded component $U$ of the Fatou set of $P_a$. Since $a \in \partial \mathcal{M}$, $U$ is not an attracting component. Since $a \in U$, there are no Siegel disks. Hence $U$ is in the basin of attraction of a parabolic cycle in $\mathcal{J}_a$.

Suppose $z_j \in \mathcal{R}(\theta, \mathcal{K}_a)$ clusters at $q_0 \in \mathcal{J}_a$. Arguing as before, this time in dynamic space, we see that $q_0$ must satisfy $P_a^{\ell+k}(q_0) = P_a^k(q_0)$. Since the solutions of this equation form a finite set, the ray must actually terminate at $q_0$. Then $\mathcal{R}(2^k\theta, \mathcal{K}_a)$ terminates at the point $q = P_a^k(q_0)$, and $q$ is a periodic point with period dividing $\ell$. Our first goal is to show that $q$ is a parabolic periodic point.

Suppose not. Since there is at most one attracting or neutral cycle, the cycle of $q$ is repelling. As $c_j \to a$, $\mathcal{R}(\theta, \mathcal{J}_{c_j}) \to \mathcal{R}(\theta, \mathcal{K}_a)$ in the sense that every compact part of $\mathcal{R}(\theta, \mathcal{K}_a)$ is approached by $\mathcal{R}(\theta, \mathcal{J}_{c_j})$. The appropriate branch of $P_a^{-\ell}$ has an attracting fixed point at $q$, so there is a disk $\Delta(q, \varepsilon)$ in the basin of attraction of $q$ mapped into a disk of smaller radius by $P_a^{-\ell}$. By the implicit function theorem (see Lemma III.2.5), $P_c^{-\ell}$ has an attracting fixed point $q(c)$ near $q = q(a)$ for all $c$ near $a$, whose basin of attraction includes

$\Delta(q, \varepsilon)$ and which maps $\Delta(q, \varepsilon)$ into a compact subset of itself. For $c$ near $a$ the ray $\mathcal{R}(2^k\theta, \mathcal{J}_c)$ remains in the disk $\Delta(q, \varepsilon)$ for a long parameter interval, and the ray is invariant under $P_c^\ell$, so the ray must terminate at $q(c)$. Thus for $c$ near $a$ the rays $\mathcal{R}(2^k\theta, \mathcal{J}_c)$ are uniformly bounded away from the parabolic cycle. On the other hand, $c_j$ belongs to $\mathcal{R}(\theta, \mathcal{J}_{c_j})$ and $c_j \to a$, so $P_{c_j}^{n\ell+k}(c_j)$ belongs to $\mathcal{R}(2^k\theta, \mathcal{J}_{c_j})$ and converges to $P_a^{n\ell+k}(a)$. Consequently $P_a^{n\ell+k}(a)$ is bounded away from the parabolic cycle. This contradicts the fact that $a$ is in the basin of attraction of the cycle. We conclude that $q$ is a parabolic periodic point.

The rays $\mathcal{R}(2^{j+k}\theta, \mathcal{K}_a)$ terminate at $P_a^j(q)$, so the condition on $\theta$ gives $P_a^\ell(q) = q$. Since $P_a^\ell$ leaves $\mathcal{R}(2^k\theta, \mathcal{K}_a)$ invariant, it does not rotate the petals at $q$, and $(P_a^\ell)'(q) = 1$. Now the solution set of the pair of equations $P_c^\ell(z) = z$, $(P_c^\ell)'(z) = 1$, is a complex analytic variety in $(z, c)$-space. For each fixed $z$, the variety is bounded (in fact, $c \in \mathcal{M}$), so that $c$ is locally constant on the variety. For fixed $c$, there are only finitely many solutions $z$. It follows that the variety is finite, and there are only finitely many possibilities for $a$.

Since there are also only finitely many solutions of $P_a^{\ell+k}(a) = P_a^k(a)$, there are only finitely many possibilities for the cluster points $a$ of the ray $\mathcal{R}(\theta, \mathcal{M})$. It follows that the ray terminates, and the first statement of the theorem is proved. It remains to show in the parabolic case that $\theta$ has odd denominator. For this, we begin by showing that the terminal point $q_0$ of the ray $\mathcal{R}(\theta, \mathcal{K}_a)$ is actually a parabolic periodic point, and then we will argue that $\theta$ is periodic.

We parametrize $\mathcal{R}(\theta, \mathcal{J}_{c_j})$ by $z_j(t) = \varphi_{c_j}^{-1}(e^t e^{2\pi i\theta})$, $t > 0$, so that $G_{c_j}(z_j(t)) = t$. Let $q_1$ be the first point in the parabolic cycle that is a cluster point of the rays $\mathcal{R}(\theta, \mathcal{J}_{c_j})$ as they come in from $\infty$. By this we mean that for $\varepsilon > 0$ small and $j$ large, there is $t_j > 0$ such that $|z(t_j) - q_1| = \varepsilon$, while the distance from $z_j(t)$ to other points of the parabolic cycle is bounded away from $0$ for all $t > t_j$, uniformly in $j$. There is then $\beta > 0$ such that $|P_{c_j}^\ell(z_j(t)) - z_j(t)| \geq \beta$ for $t \geq t_j$. Consider $g_j(\zeta) = \varphi_{c_j}^{-1}(1/\zeta)$, which is univalent for $|\zeta| < \exp(-G_{c_j}(0)) = \rho_j$. Note that $t_j > G_{c_j}(c_j) = 2G_{c_j}(0)$ for $j$ large, so $|g_j^{-1}(z_j(t))| = e^{-t} < \rho_j^2$ for $t \geq t_j$, and the points $g_j^{-1}(z_j(t))$, $t \geq t_j$, are uniformly bounded in the hyperbolic metric of the disk $\Delta(0, \rho_j)$ on which $g_j$ is univalent. Now if $t \geq t_j$ and $g_j^{-1}(z_j(t)) = re^{2\pi i\theta}$, then $g_j^{-1}(P_{c_j}^\ell(z_j(t))) = r^{2^\ell}e^{2\pi i\theta}$ is at a uniformly bounded hyperbolic distance from $g_j^{-1}(z_j(t))$ in the disk $\Delta(0, \rho_j)$.

Since $|g_j(re^{2\pi i\theta}) - g_j(r^{2^\ell}e^{2\pi i\theta})| \geq \beta$, we see from the distortion theorem scaled to $\Delta(0, \rho_j)$ that for some small $\delta > 0$ the image of $\Delta(0, \rho_j)$ under $g_j$ contains a disk centered at $\varphi_{c_j}(re^{2\pi i\theta})$ of radius $\delta$. Here $\delta$ depends on $\varepsilon$ and $\beta$ but not on $j$. Thus for $j$ large, $\mathcal{R}(\theta, \mathcal{J}_{c_j})$ stays at a uniform distance $\delta$ from $\mathcal{J}_{c_j}$, and hence at a uniform distance from $\mathcal{J}_a$, until entering an $\varepsilon$-neighborhood of $q_1$. It follows that $q_1$ must coincide with the terminal point $q_0$ of $\mathcal{R}(\theta, \mathcal{K}_a)$. Thus $q_0$ is a parabolic periodic point. (With a little additional effort, we could establish that $q_0 = q$. The limit of the rays $\mathcal{R}(c_j, \mathcal{J}_{c_j})$ is a ray which traverses $\mathcal{R}(\theta, \mathcal{K}_a)$ to the parabolic point $q$, then makes a sharp turn depending on the number of petals, and continues upstream along the flow lines in the basin of attraction of $q$ to the critical value $a$ and beyond.)

For $j \geq 1$, the rays $\mathcal{R}(2^{j\ell}\theta, \mathcal{K}_a)$ terminate at $P_a^{j\ell}(q_0)$. Take $n$ so that $n\ell \geq k$. The preperiodicity condition shows that these rays coincide for $j \geq n$, and in particular they terminate at the same point $P_a^\ell(q_0)$. Since $P_a^\ell$ maps this family of rays to itself, and since $P_a^\ell$ is conformal near $q_0$, there can be only one ray in the family. Hence $\mathcal{R}(\theta, \mathcal{K}_a)$ is invariant under $P_a^\ell$. We conclude that $2^\ell\theta \equiv \theta$, and $\theta$ is periodic. $\square$

Note in the parabolic case that the period $\ell$ of $\theta$ coincides with the number of petals of the parabolic cycle of $P_a$. Indeed, the proof shows that $\ell$ divides the total number $N$ of petals ($=$ number cusps). By Theorem III.2.3, each petal contains a critical point of $P_a^N$, so that the orbit of $a$ hits each petal. Thus $P_a$ permutes the cusps cyclically, and $\ell = N$.

Theorem VII.2.2 shows that there are a finite number of repelling arms for $P_a^\ell$ terminating at any parabolic periodic point $q$, and these arms fill out the unbounded component of the Fatou set near $q$. Using Lindelöf's theorem, we deduce that through each repelling arm travels exactly one external ray of $\mathcal{K}_a$ that terminates at $q$. Now $P_a^\ell$ leaves each cusp between consecutive petals invariant, it permutes the rays terminating at $q$, and it preserves the circular order of the rays. It follows that each ray is invariant under $P_a^\ell$, and furthermore the same number of external rays terminate through each cusp at each point of the cycle. It turns out that at most two rays terminate through each cusp. More precise information will be given in Theorem 7.1.

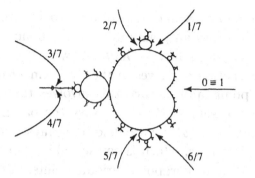

FIGURE 9. External rays terminating at period-three parabolic points of $\mathcal{M}$.

EXAMPLE. Consider the external angles whose binary representations have period three. There are exactly six of them, $1/7, ..., 6/7$, and each corresponding ray terminates at a parabolic point $a \in \partial \mathcal{M}$ for which there is a cycle of attracting petals of length three (see Figure 9). Either $P_a$ has a parabolic fixed point with multiplier $e^{\pm 2\pi i/3}$, or $P_a$ has a parabolic cycle of period three and multiplier 1. The equations give only three possibilities for $a$, and these are the roots of the three hyperbolic components corresponding to attracting three-cycles. The rays $\mathcal{R}(1/7, \mathcal{M})$ and $\mathcal{R}(2/7, \mathcal{M})$ terminate at the point of tangency of the main cardioid and the big bulb at the top of the main cardioid. The two rays approach their terminal point from opposite directions and cut the Mandelbrot set into two pieces. Their reflections in the lower half-plane are the rays $\mathcal{R}(5/7, \mathcal{M})$ and $\mathcal{R}(6/7, \mathcal{M})$. The rays $\mathcal{R}(3/7, \mathcal{M})$ and $\mathcal{R}(4/7, \mathcal{M})$ approach the root $r$ of the third component from above and below the real axis. (See the example in Section 2.) In the dynamic plane, the two rays $\mathcal{R}(3/7, \mathcal{K}_r)$ and $\mathcal{R}(4/7, \mathcal{K}_r)$ approach a parabolic periodic point $q$ on the real axis of period three from the respective half-planes. There is only one attracting petal at $q$. This provides the simplest example for which there is more than one ray terminating through the same cusp at a parabolic periodic point.

## 6.  Misiurewicz Points

We turn now to a discussion of the converse to Theorem 5.2, that all Misiurewicz points and all points in $\mathcal{M}$ corresponding to parabolic cycles are the endpoints of rays $\mathcal{R}(\theta, \mathcal{M})$ with $\theta$ rational. This involves the construction of curves in the complement of $\mathcal{M}$ that terminate at the point, and in turn this involves the deformation of paths in the Fatou sets $\mathcal{F}(P_c)$ as we vary the parameter $c$. In the case of Misiurewicz points, we are deforming rays that terminate at repelling periodic points, which is considerably easier than deforming rays terminating at parabolic periodic points. (Viewed this way, repelling points are attractive.) In this section, we give a reasonably complete treatment of the repelling (Misiurewicz) case. In the final section, we content ourselves with describing the results and indicating a deformation method for the parabolic case.

We begin with a more detailed description of the Julia set corresponding to a Misiurewicz point. Let $a$ be a Misiurewicz point, with critical orbit

$$0 \to a \to P_a(a) \to \cdots \to P_a^k(a) \to \cdots \to P_a^{m+k}(a) = P_a^k(a),$$

so that $a$ enters the cycle at the $k$th iteration and the cycle has period $m$. Since $P_a$ is subhyperbolic, the cycle is repelling. Since the only preimage of $a$ is 0, the critical value $a$ does not belong to the cycle, and $k \geq 1$. By Theorem V.4.2 the Julia set $\mathcal{J}_a = \mathcal{K}_a$ is a dendrite, and the inverse of the conjugating map $\varphi_a$ extends continuously to map the unit circle onto $\mathcal{J}_a$. Thus every external ray $\mathcal{R}(\theta, \mathcal{K}_a)$ terminates at a point of $\mathcal{K}_a$, and every point of $\mathcal{K}_a$ has at least one external angle.

By Theorem VII.2.2, there are a finite number $N$ of repelling arms at $P_a^k(a)$, and these fill out the Fatou set near $P_a^k(a)$. From Lindelöf's theorem (Theorem I.2.2) we see that through each repelling arm there is exactly one external ray terminating at $P_a^k(a)$. The appropriate inverse branch of $P^{-k}$ maps these onto $N$ arms of the Fatou set that fill out the Fatou set near $a$, each of which has exactly one external ray terminating at $a$. Let the external angles of $\mathcal{K}_a$ at $a$ be $\theta_1, ..., \theta_N$. The successive images of the $i$th arm under $P_a$ then have external angles $2\theta_i, 4\theta_i, \ldots$, all reduced mod 1. The external angles are circularly ordered, and since $P_a$ preserves the order, so does the doubling operation on the external angles. Thus if one of the repelling arms at $P_a^k(a)$ has period $\ell$, they all do, and consequently

$2^{k+\ell}\theta_i \equiv 2^k\theta_i$ for $1 \leq i \leq N$, that is, each of the $\theta_i$'s is strictly preperiodic, $k$ is the first integer such that $2^k\theta_i$ is periodic, and the periods are all the same, equal to the period $\ell$ of the repelling arms.

EXAMPLE. Consider the ray $\mathcal{R}(1/6, \mathcal{M})$, which terminates at some $a \in \mathcal{M}$ with $\operatorname{Im} a \geq 0$. The binary representation $1/6 \sim (0,0,1,0,1, 0,1,...)$ is strictly preperiodic. From this representation we deduce that $P_a^3(a) = P_a(a)$. Now

$$P_c^3(c) - P_c(c) = c^3(c+2)(c+1)^2(c^2+1).$$

The roots 0 and $-1$ corresponding to superattracting cycles can be discarded. The ray cannot terminate at $-2$, which is at the left tip of $\mathcal{M}$. The only possibility is that $\mathcal{R}(1/6, \mathcal{M})$ terminates at $i$. The arms at $P_i(i) = i - 1$ have period two. Since $i - 1 \to -i$, there can be only one arm at each of the points $i - 1$ and $-i$, so these points lie at tips of the dendrite, as does $a = i$. See the figure in Section V.4.

EXAMPLE. Let $a$ be a Misiurewicz point such that $P_a^2(a)$ is a fixed point for $P_a$, that is, such that $P_a^3(a) = P_a^2(a)$. Since

$$P_c^3(c) - P_c^2(c) = c^8 + 4c^7 + 6c^6 + 6c^5 + 4c^4 = c^4(c+2)(c^3+2c^2+2c+2),$$

we find upon discarding the superattracting cycle exactly four possibilities. The root $a = -2$ corresponds to the critical orbit $0 \to -2 \to 2 \to 2 \to \cdots$, so $P_a(a)$ is already a fixed point. The cubic factor has one real root $r_0 \approx -1.54369$ and two complex roots. One can verify that the rays with external angles

$$\frac{5}{12} \sim (0,1,1,0,1,0,1,...), \qquad \frac{7}{12} \sim (1,0,0,1,0,1,0,...),$$

correspond to a Misiurewicz point $a$ on the real axis with two repelling arms, one above and one below the real axis, which are interchanged by $P_a$, and that this $a$-value coincides with $r_0$. The complex roots correspond to rays with external angles $1/4 \sim (0,0,1,1,1,...)$ and $3/4 \sim (1,1,0,0,0,...)$.

Now we wish to transfer information about the Julia set at $a$ to the parameter plane, to obtain information about $\mathcal{M}$ at $a$. We continue with the notation above.

Consider again the polynomial $Q(z,c) = P_c^m(z) - z$ in $z$ and $c$. Denote by $z_0 = z_0(a)$ the first point $P_a^k(a)$ of the repelling cycle, and denote by $\lambda(a)$ the multiplier of the cycle, so that $|\lambda(a)| > 1$. Then $z_0$ is a fixed point of $P_a^m$, so $Q(z_0, a) = 0$. As before, the implicit function theorem guarantees that the equation $Q(z, c) = 0$ has a unique solution $z_0(c)$ near $z_0(a)$ for $c$ near $a$, which depends analytically on $c$. The point $z_0(c)$ is a repelling periodic point of period $m$ for $P_c$, with multiplier $\lambda(c)$ depending analytically on $c$.

Let $h_c$ map a disk $\{|\zeta| < \delta\}$ to a neighborhood of $z_0(c)$, so that $h_c$ is a conjugation of $P_c^m$ and multiplication by $\lambda(c)$:

$$h_c(\lambda(c)\zeta) = P_c^m(h_c(\zeta)), \quad \lambda(c)h_c^{-1}(z) = h_c^{-1}(P_c^m(z)), \qquad |\zeta| < \delta.$$

We normalize $h_c$ so that $h_c'(0) = 1$, and then $h_c$ depends analytically on the parameter $c$.

Fix an external angle $\theta_j$ of $\mathcal{K}_a$ at $a$, and let $\phi \equiv 2^k \theta_j$, so that $\mathcal{R}(\phi, \mathcal{K}_a)$ terminates at $z_0(a)$. We consider the rays $\mathcal{R}(\phi, \mathcal{K}_c)$ for $c$ near $a$. These vary analytically with $c$ outside any neighborhood of $z_0(a)$. Near $z_0(c)$ the rays are invariant under $P_c^{-m}$, and the relation

$$P_c^{-m}(z) = h_c\left(\frac{1}{\lambda(c)}h_c^{-1}(z)\right)$$

shows that the rays terminate at $z_0(c)$, uniformly for $c$ near $a$. For fixed $t > 0$, let $z_t(c)$ be the point on the ray $\mathcal{R}(\phi, \mathcal{K}_c)$ that satisfies $G_c(z_t(c)) = t$, that is,

$$\varphi_c(z_t(c)) = e^t e^{2\pi i \phi}, \qquad t > 0.$$

Then $z_t(c)$ depends continuously on $t$ and $c$, $z_t(c)$ depends analytically on $c$ for each fixed $t$, and $z_t(c)$ tends to $z_0(c)$ uniformly for $c$ near $a$ as $t$ decreases to 0.

Now $z_0(c)$ does not coincide identically with $P_c^k(c)$ near $a$, or we would have $P_c^{m+k}(c) = P_c^k(c)$ identically. Thus for some $\nu \geq 1$ and $A_\nu \neq 0$, we have

$$z_0(c) - P_c^k(c) = A_\nu(c - a)^\nu + \mathcal{O}(|c - a|^{\nu+1}).$$

For $t > 0$ small, the equation $z_t(c) - P_c^k(c) = 0$ has $\nu$ solutions $c_1(t), ..., c_\nu(t)$ counting multiplicities, depending continuously on $t$ and tending to $a$ as $t \to 0$. In particular, $P_{c_i(t)}^k(c_i(t)) \in \mathcal{R}(\phi, \mathcal{K}_{c_i(t)})$. By applying the appropriate branch of $P_{c_i(t)}^{-k}$, we obtain $c_i(t) \in \mathcal{R}(\theta_j, \mathcal{K}_{c_i(t)})$. Hence $c_i(t) \in \mathcal{R}(\theta_j, \mathcal{M})$, and it follows that $\mathcal{R}(\theta_j, \mathcal{M})$ terminates at $a$. We have proved the following.

THEOREM 6.1. *If $a$ is a Misiurewicz point, then there are a finite number of external rays $\mathcal{R}(\theta_j, \mathcal{K}_a)$ of $\mathcal{K}_a$ that terminate at $a$. Each external angle $\theta_j$ of $\mathcal{K}_a$ at $a$ is rational and strictly preperiodic. Moreover, each ray $\mathcal{R}(\theta_j, \mathcal{M})$ of the Mandelbrot set terminates at $a \in \mathcal{M}$. In particular, every Misiurewicz point belongs to $\partial \mathcal{M}$ and is the terminal point of $\mathcal{R}(\theta, \mathcal{M})$ for some strictly preperiodic $\theta$.*

We wish to analyze the situation at the point $a$ more carefully. We aim to show that the Julia set $\mathcal{J}_a$ is asymptotically similar at $a$ to the Mandelbrot set $\mathcal{M}$, and in particular if $\overline{\mathbb{C}} \backslash \mathcal{J}_a$ has $N$ arms at $a$, then $\overline{\mathbb{C}} \backslash \mathcal{M}$ also has $N$ arms at $a$. We begin by showing that the multiplicity $\nu$ above is 1.

LEMMA 6.2. *With the notation above, $z_0(c) - P_c^k(c)$ has a simple zero at $a$, as does $P_c^{m+k}(c) - P_c^k(c)$.*

*Proof.* From

$$\varphi_{c_i(t)}(P_{c_i(t)}^k(c_i(t))) = \varphi_{c_i(t)}(z_t(c_i(t))) = e^t e^{2\pi i \phi},$$

we have, since $c_i(t) \in \mathcal{R}(\theta_j, \mathcal{K}_{c_i(t)})$, that

$$\varphi_{c_i(t)}(c_i(t)) = e^{t/2^k} e^{2\pi i \theta_j}.$$

Since $\Phi$ is one-to-one, the $c_i(t)$'s coincide, call their common value $c(t)$. Thus

$$z_t(c) - P_c^k(c) = \mathcal{O}(|c - c(t)|^\nu), \qquad c \to c(t).$$

Now $\varphi_c(z) = e^t e^{2\pi i \phi} + \mathcal{O}(|z - z_t(c)|)$, so

$$\Phi(c)^{2^k} = \varphi_c(P_c^k(c)) = e^t e^{2\pi i \phi} + \mathcal{O}(|c - c(t)|^\nu), \qquad c \to c(t).$$

But $\Phi(c)^{2^k}$ has nonzero derivative at $c = c(t)$. We conclude that $\nu = 1$, and $z_0(c) - P_c^k(c)$ has a simple zero at $a$. We calculate

$$\frac{d}{dc}\left[P_c^{m+k}(c) - P_c^k(c)\right]_{c=a} = (\lambda(a) - 1)\frac{d}{dc}\left[P_c^k(c) - z_0(c)\right]_{c=a},$$

so the second statement is equivalent to the first. $\square$

For $c$ near $a$, define

$$Z(c) = h_a \circ h_c^{-1}(P_c^k(c)) = z_0(a) + \mathcal{O}(|c - a|).$$

Since $h'_c(0) = 1$, we have $h_c^{-1}(P_c^k(c)) = P_c^k(c) - z_0(c) + \mathcal{O}(|P_c^k(c) - z_0(c)|^2)$. From the preceding lemma, $h_c^{-1}(P_c^k(c))$ has a simple zero at $a$, so $Z(c)$ is univalent near $a$.

We wish to compare the orbits $P_c^{jm+k}(c)$ and $P_a^{jm}(Z(c))$, $j \geq 1$. We compare the orbits in conjugation space, where the dynamics are simply multiplication by $\lambda(c)$ and $\lambda(a)$, respectively. We have defined $Z(c)$ so that the starting points $\zeta_0 = h_c^{-1}(P_c^k(c)) = h_a^{-1}(Z(c))$ are the same, and at time $jm$ we have $\lambda(c)^j\zeta_0$ and $\lambda(a)^j\zeta_0$, respectively. We iterate until we are about to leave a disk of fixed size, say $\{|\zeta| \leq C_1\}$, at the $J$th iteration, where $J = J(c)$. For $\zeta_0$ we have the estimate

$$|\zeta_0| = |h_a^{-1}(Z(c))| \geq C_0|c - a|,$$

since $h_a^{-1}$ is analytic and $h_a^{-1}(z_0(a)) = 0$. Thus $C_0|c-a||\lambda(a)|^J \leq C_1$, and

$$J \leq C_2 \log \frac{1}{|c - a|}.$$

Since $\lambda(c)$ is an analytic function of $c$, we have $\lambda(c)/\lambda(a) = 1 + \mathcal{O}(|c - a|)$, and thus

$$
\begin{aligned}
|\lambda(c)^J\zeta_0 - \lambda(a)^J\zeta_0| &= |\zeta_0||\lambda(a)|^J\left|1 - \left(\frac{\lambda(c)}{\lambda(a)}\right)^J\right| \\
&\leq C_1|1 - (1 + \mathcal{O}(|c - a|))^J| \\
&\leq C_3|c - a| \log \frac{1}{|c - a|}.
\end{aligned}
$$

Going back to the $z$-plane, we obtain an estimate of the form

$$|P_c^{jm+k}(c) - P_a^{jm}(Z(c))| \leq C|c - a| \log \frac{1}{|c - a|} \tag{6.1}$$

so long as the iterates remain inside a fixed disk centered at $z_0(a)$.

Fix a repelling arm for $P_a^m$ at $P_a^k(a) = z_0(a)$, call it $S$. Parametrize $S$ by $\xi = \varphi(z)$, so $\alpha < \operatorname{Im} \xi < \beta$ and $P_a^m$ is translation by 1 in the $\xi$-coordinate. Let $S_\varepsilon$ be the subset of $S$ represented by parameter values $\alpha + \varepsilon \leq \operatorname{Im} \xi \leq \beta - \varepsilon$, $\operatorname{Re} \xi \leq \gamma$. Let $T_\varepsilon$ be a period rectangle in $S_\varepsilon$ of the form $\{z \in S_\varepsilon : \gamma - 1 \leq \operatorname{Re} \xi \leq \gamma\}$, where $\gamma$ is large negative so that $T_\varepsilon$ is close to $a$. Let $V_\varepsilon$ be a neighborhood of $T_\varepsilon$ that is compact in $S$. Choose $N_0$ so that all points in $V_\varepsilon$ are iterated to $|z| > 4$ in at most $N_0$ steps by $P_c$ for all $c$ near $a$.

We claim that if $c$ is near $a$ and $Z(c) \in S_\varepsilon$, then $c \notin M$. Indeed the estimate (6.1) shows that when eventually the iterates $P_a^{jm}(Z(c))$ hit

$T_\varepsilon$, the corresponding iterates $P_c^{jm+k}(c)$ hit $V_\varepsilon$, from whence in $N_0$ more steps they move to $|z| \geq 4$ and then beyond towards $\infty$. Thus the image of $S$ under $Z^{-1}$ is asymptotic to an "*arm*" of $\overline{\mathbf{C}} \backslash \mathcal{M}$ at $a$.

Now suppose that $\phi = 2^k \theta_j$ is the external angle of the ray terminating at $z_0(a)$ through $S$. Since the ray $\mathcal{R}(\phi, \mathcal{K}_a)$ is periodic in $\xi$-space, it lies in $S_{2\varepsilon}$ for $\varepsilon > 0$ small. For $c(t)$ near $a$ the ray $\mathcal{R}(\phi, K_{c(t)})$ passes through $T_\varepsilon$. Hence for $c(t)$ sufficiently close to $a$, we obtain from (6.1) that $P_a^{jm}(Z(c(t))) \in V_\varepsilon$, and consequently $Z(c(t)) \in S$. Thus the ray $\mathcal{R}(\theta_j, \mathcal{M})$ terminates at $a$ through the set $Z^{-1}(S)$. The argument shows in fact that the hyperbolic distance between points of $\mathcal{R}(\theta_j, \mathcal{K}_a)$ and $\mathcal{R}(\theta_j, \mathcal{M})$, with respect to $\overline{\mathbf{C}} \backslash \mathcal{M}$, tends to 0 as the rays tend to their common terminal point $a$. We have proved the following version of a theorem of Tan Lei (Exposé No. XXIII of [DH2]).

THEOREM 6.3. *Let $a \in \mathcal{M}$ be a Misiurewicz point. Let $\theta_1, ..., \theta_N$ be the external angles of $\mathcal{J}_a$ at $a$, and let $L_j$ be the arm of $\mathcal{F}_a$ containing $\mathcal{R}(\theta_j, \mathcal{J}_a)$, so that $S_j = P_a^k(L_j)$ is a repelling arm at $P_a^k(a)$ as above, and $L_1, ..., L_N$ fill out the Fatou set near $a$. Parametrize $L_j$ by $\zeta$ in a strip $a_j < \operatorname{Im} \zeta < b_j$, $\operatorname{Re} \zeta \to -\infty$, and for $\varepsilon > 0$ small let $L_{j,\varepsilon}$ be the subset of the arm corresponding to $a_j + \varepsilon < \operatorname{Im} \zeta < b_j - \varepsilon$. Let $Z$ be as above, and define the conformal map $g = Z^{-1} \circ P_a^k$, so that $g(a) = a$ and $g'(a) \neq 0$. Then $g(L_{j,\varepsilon})$ eventually lies in $\mathbf{C} \backslash \mathcal{M}$, and the ray $\mathcal{R}(\theta_j, \mathcal{M})$ terminates at $a$ through $g(L_{j,\varepsilon})$.*

One can note that computer-generated pictures of certain pieces of the Mandelbrot set in parameter space look strikingly similar to pieces of Julia sets in dynamic space, with swirling arms configured in the same pattern. See particularly Figures 4.22 and 4.23 on pages 204-206 of [Pe], and also the figures in [Ta]. An important step in the proof of Shishikura [Sh2] that the boundary of the Mandelbrot set has Hausdorff dimension two hinges on the local similarity of the Mandelbrot set and Julia sets.

## 7.   Parabolic Points

As noted before, the analysis of parabolic points of $\mathcal{M}$ is more difficult than that of Misiurewicz points. We summarize the situation in Theorems 7.1 and 7.2, without supplying detailed proofs. For a complete discussion, see [**DH2**].

THEOREM 7.1. *Suppose $P_a$ has a parabolic cycle of length $m$ with multiplier $\lambda$. If $\lambda \neq 1$, then between any pair of consecutive petals at any point $q$ of the cycle, there is exactly one external ray of $\mathcal{K}_a$ that terminates at $q$, and these are permuted cyclically by $P_a$. If $\lambda = 1$ and $a \neq 1/4$, then there are two external rays terminating at any point of the cycle, and the set of pairs is permuted cyclically by $P_a$. In any case, the corresponding external angles are all rational and periodic, with period $\ell = mN$, where $N$ is the number of petals at $q$.*

The external rays of $\mathcal{K}_a$ in dynamic space are again related to those of the Mandelbrot set in parameter space. By Theorem 5.2, every external ray $\mathcal{R}(\theta, \mathcal{M})$ of the Mandelbrot set with periodic $\theta$ terminates at a parabolic point $a \in \mathcal{M}$. Conversely, every parabolic point $a$ is the terminal point of such a ray.

THEOREM 7.2. *Suppose that $a \in \mathcal{M}$ is such that $P_a$ has a parabolic periodic point. Except in the trivial case $a = 1/4$ (corresponding to a parabolic fixed point, with one ray $\mathcal{R}(0, \mathcal{M})$ terminating at $a$), there are exactly two rays $\mathcal{R}(\theta_1, \mathcal{M})$, $\mathcal{R}(\theta_2, \mathcal{M})$ terminating at $a$. The external angles $\theta_1, \theta_2$ coincide with the angles of the external rays of $\mathcal{K}_a$ tangent to and on either side of the petal containing $a$.*

We conclude by focusing on one part of the proofs of Theorems 7.1 and 7.2, the existence of an external ray of $\mathcal{M}$ that terminates at the parabolic point $a$. This will serve to shed light on the phenomena behind the results and what kind of arguments go into the proof.

For $\ell$ the period of a petal, as above, we expand

$$P_c^\ell(z) = \sum \alpha_j(c)(z - q)^j,$$

where $\alpha_0(a) = 0$ and $\alpha_1(a) = 1$. If the flower at $q$ has $N$ petals, then $\alpha_2(a) = \cdots = \alpha_N(a) = 0$ and $\alpha_{N+1}(a) \neq 0$. We will consider only the special case $N = 1$; that is, we assume there is only one petal at $q$. By moving the point near $q$ where $(P_c^\ell)' = 1$ to 0 and scaling, we can then assume that our family $P_c^\ell(z)$ has the expansion

$$f(z, c) = \gamma(c)^2 + z + z^2 + B(c)z^3 + \mathcal{O}(z^4) = \gamma(c)^2 + z + F(z, c).$$

Here $\gamma(c)^2$ is analytic and not identically zero, $\gamma(a)^2 = 0$, and $\gamma(c)$ is the principal branch of the square root. The point $z = 0$ is a parabolic fixed point for $f(z, a)$, which belongs to the Julia set and also to the boundary of the unbounded component of the Fatou set of $f(z, a)$. Let $z_0^*$ belong to this component and be very close to $z = 0$. We denote by $\Gamma^*$ a compact subset of this component containing $z_0^*$, which will be specified later. The target set $\Gamma^*$ will be situated near $0$ in the repelling arm of the Fatou set containing $z_0^*$, and it will be large enough so that the following statement holds.

*Main deformation construction.* Let $z_0(c)$ be any function analytic near $a$ so that $z_0(a)$ is in the parabolic petal of $0$. Then there is a curve $\Gamma$ in parameter space, terminating at $a$, so that for $c \in \Gamma$, $c \neq a$, there exists $n = n(c)$ such that $f^n(z_0(c), c) \in \Gamma^*$.

In our application, $z_0(c)$ is the point corresponding under the conjugation to the critical point $0$. Thus if $c \in \Gamma$ is near $a$, the critical point is iterated first to a point in a fixed compact subset of the unbounded component of the Fatou set of $f(z, a)$ and then, by continuity, to $\infty$. Hence $\Gamma$ is in the complement of the Mandelbrot set. By Lindelöf's theorem, this implies there exists a ray $\mathcal{R}(\theta, \mathcal{M})$ which terminates at $a$.

Since the iterates of $z_0(a)$ eventually approach $0$ through a narrow cusp tangent to the negative real axis, we focus on initial values $z_0(c)$ that lie in some fixed domain $\Omega$ containing $-\varepsilon$, so that $x \approx -\varepsilon$ and $|y| \ll \varepsilon$ on $\Omega$. We study parameter values $c$ for which $\gamma(c) = \alpha + i\beta$ where $\alpha > 0$ and $|\beta| \ll \alpha$. In fact, we shall see we can take $|\beta| \leq C\alpha^2$ or $|\beta| \leq C\alpha^2 \log(1/\alpha)$, depending on whether $\operatorname{Im} B(a)$ is $0$ or not. We regard $\alpha$ as the main parameter, and we seek parameter values $c = c(\alpha)$ so that the parabolic fixed point at $0$ splits into two fixed points, above and below the real axis, opening a window through which the iterates pass from $\Omega$ in the left half-plane to $\Gamma^*$ in the right. As the window narrows, the number of iterations required to pass through it becomes large.

Now the iteration

$$z_{n+1} = z_n + \gamma(c)^2 + F(z_n, c) \tag{7.1}$$

can be approximated by the flow

$$\frac{dz(t)}{dt} = \gamma(c)^2 + F(z, c), \qquad z(0) = z_0 = x_0 + iy_0 \in \Omega. \tag{7.2}$$

The equation (7.2) is easily solved for $t$,

$$t = t(z, c) = \int_{z_0}^{z} \frac{d\zeta}{\gamma(c)^2 + F(\zeta, c)}.$$

We are interested in the point $iy$ where the solution curve meets the imaginary axis. The corresponding parameter value $t$ is given by

$$t = \int_{z_0}^{-\varepsilon} + \int_{-\varepsilon}^{0} + \int_{0}^{iy} \frac{d\zeta}{\gamma(c)^2 + F(\zeta, c)}.$$

The first integral is analytic in $c$. In the second, we expand the integrand $1/[\gamma(c)^2 + F(\zeta, c)]$ as

$$\frac{1}{\gamma(c)^2 + \zeta^2} - \frac{B(c)\zeta}{\gamma(c)^2 + \zeta^2} + \frac{B(c)\gamma(c)^2\zeta}{(\gamma(c)^2 + \zeta^2)^2} + \frac{\mathcal{O}(\zeta^4)}{(\gamma(c)^2 + \zeta^2)^2},$$

and we integrate, thereby obtaining

$$t = A(z_0, c) + \frac{\pi}{2\gamma(c)} + B(c)\log\frac{1}{\gamma(c)} + i\int_{0}^{y} \frac{ds}{\gamma(c)^2 - s^2 + \mathcal{O}(s^3)},$$

where $A$ is analytic and uniformly bounded for $\operatorname{Re}\gamma(c)^2 > 0$. Note also that for $\gamma = \alpha + i\beta$ as above, $|dB/d\gamma| < C|\gamma|^{\delta-1}$ for some $\delta > 0$, and $|dA/d\gamma| < C/|\gamma|$.

For fixed $\alpha > 0$, the condition on $\beta$ that the curve pass through the origin is that

$$\operatorname{Im} A(z_0, c) - \frac{\pi\beta}{2|\gamma(c)|^2} + \frac{\operatorname{Re} B(c)}{2}\operatorname{Arg}\frac{1}{\gamma(c)^2} + \frac{\operatorname{Im} B(c)}{2}\log\frac{1}{|\gamma(c)|^2} = 0.$$

Using $|\gamma(c)|^2 \approx \alpha^2$ and $\operatorname{Arg}\gamma(c)^2 \approx 2\beta/\alpha$, we see that the expression on the left moves continuously from positive to negative as $\beta$ increases between the limits $\pm C\alpha^2\log(1/\alpha)$. Hence it is zero for some $\beta \approx (2/\pi)\operatorname{Im} B(a)\alpha^2\log(1/\alpha)$. For $|\beta| < C\alpha^2\log(1/\alpha)$, the curve crosses the imaginary axis at $iy$, where

$$y \approx \frac{\pi}{2}\beta - \alpha^2\operatorname{Im} A(z_0, c) + \alpha\beta\operatorname{Re} B(c) - \alpha^2\log(1/\alpha)\operatorname{Im} B(c).$$

Using the estimates mentioned above, we obtain

$$\frac{dy}{d\beta} = \frac{\pi}{2} + \mathcal{O}(\alpha).$$

In particular, the $y$-value of the crossing point is an *increasing* function of $\beta$.

Now, let $\tilde{f}(z,c)$ determine a similar flow in the opposite direction, starting at $\tilde{z}_0$ and approaching the imaginary axis from the right. We assume $\tilde{f}$ has the same leading term but with opposite sign:

$$\tilde{f}(z,c) = -\gamma(c)^2 + z - z^2 + \tilde{B}(c)z^3 + \mathcal{O}(z^4) = z - (\gamma(c)^2 + \tilde{F}(z,c)).$$

We conjugate $z \to -z$ and use the same argument. In this case, the $y$-value of the crossing point on the imaginary axis is a *decreasing* function of $\beta$. Thus we conclude the following.

THEOREM 7.3. *Given $z_0(c)$, $\mathrm{Re}\, z_0 < 0$ and $\tilde{z}_0(c)$, $\mathrm{Re}\, \tilde{z}_0(c) > 0$, as above, there exists a Jordan curve $\Gamma$ of parameter values $c$ terminating at $c = a$ and satisfying $\alpha > 0$ and $|\beta| < C\alpha^2 \log(1/\alpha)$, where $\gamma(c) = \alpha + i\beta$, so that the two solution curves of*

$$\frac{dz(t)}{dt} = \gamma(c)^2 + F(z,c), \qquad \mathrm{Re}\, z \le 0, \ z(0) = z_0(c),$$
$$\frac{d\tilde{z}(t)}{dt} = -\gamma(c)^2 - \tilde{F}(\tilde{z},c), \qquad \mathrm{Re}\, \tilde{z} \ge 0, \ \tilde{z}(0) = \tilde{z}_0(c),$$

*meet on the imaginary axis.*

We can also apply the analysis to $f^{-1}(z,c)$, which has the expansion

$$z - \gamma(c)^2 - (z - \gamma(c)^2)^2 - (B(c) - 2)(z - \gamma(c)^2)^3 + \mathcal{O}((z - \gamma(c)^2)^4).$$

After a conjugation $z' = z - \gamma(c)^2$, this assumes the form $\tilde{f}(z,c)$ treated above. The lines of flow of $f^{-1}(z,c)$ arrive from the right, cross over the vertical line $\mathrm{Re}\,(z - \gamma(c)^2) = 0$, and hit the imaginary axis at a crossing point that is a decreasing function of $\beta$. Thus we obtain the main deformation result, in the case of flows, with $\Gamma^*$ the single point $z_0^*$.

When we consider iterations rather than flows, the situation is quite analogous, but the technical changes are somewhat tedious. We face two problems.

(a) For the discrete iteration, we cannot expect to hit the imaginary axis, but to come within $C|\gamma|^2$. This is the reason why $\Gamma^*$ must be considered.

(b) We wish to use the flow to follow the iteration. However, the (Euler) approximation (7.2) to (7.1) is not accurate enough.

Let us deal first with the second problem. Instead of using (7.2), we consider the flow

$$\frac{dw(t)}{dt} = \gamma(c)^2 + w^2 + (B(c) - 1)w^3, \qquad (7.3)$$

which can be regarded as a (backward) Runge–Kutta approximation for (7.2).

LEMMA. *If* $z_0 = x_0 + iy_0$, *where* $-\varepsilon \le x_0 \le -\sqrt{\alpha}$ *and* $|y_0| \ll |x_0|$, *then*

$$\int_{z_0}^{z_0 + f(z_0, c)} \frac{dz}{\gamma(c)^2 + z^2 + (B(c) - 1)z^3} = 1 + \mathcal{O}(|z_0|^2).$$

*Proof.* Expand the integrand, integrate the leading terms in closed form (as before), and make the obvious estimates. $\square$

Now let $W_\alpha$ be a rectangular window centered at $-\sqrt{\alpha}$, of width (say) $3\alpha$, and of height sufficient so that the flow lines of (7.3) starting in $\Omega$ pass through $W_\alpha$. Take $\zeta_0 \in W_\alpha$ on such a flow line, and consider backward iterates $\zeta_j = f(\zeta_{j+1})$. We compare the polygonal line $P_n$ starting at $\zeta_0$ and passing through $\zeta_1, ..., \zeta_n$ to the backward flow $w(-t)$ starting at the same point $\zeta_0$. From $\zeta_{j+1}^2 = \zeta_j - \zeta_{j+1} + \mathcal{O}(\zeta_{j+1}^3)$ and $\mathrm{Re}\,(\zeta_j - \zeta_{j+1}) \sim |\zeta_j - \zeta_{j+1}|$, we obtain

$$\int_{P_n} \frac{d\zeta}{\gamma(c)^2 + \zeta^2 + (B(c) - 1)\zeta^3} = -n + \mathcal{O}(\textstyle\sum |\zeta_j|^2) = -n + \mathcal{O}(|\zeta_n|).$$

Since the integrand behaves like $1/\zeta^2$, this shows that $w(-n) = \zeta_n + \mathcal{O}(|\zeta_n|^3)$. We conclude that if $z_0 \in \Omega$, there is a point $w_0$ satisfying $|w_0 - z_0| = \mathcal{O}(\varepsilon^3)$, such that the forward flow (7.3) starting at $w_0$ follows closely the iterates of $z_0$ and meets them in $W_\alpha$.

Now let $D_\alpha$ be a similar window, centered at 0, of width (say) $3\alpha^2$ and height $C\alpha^2 \log(1/\alpha)$. The curvature of the flow lines of (7.2) is determined by

$$\frac{d^2 z}{dt^2} = (2z + 3B(c)z^2 + \cdots)\frac{dz}{dt}.$$

The curve $\text{Im}\,(2z+3B(c)z^2+\cdots) = 0$ is analytic at 0 with horizontal tangent. Above this curve, where the imaginary part is positive, the flow lines are concave up, and below it they are concave down. Thus there is a cusp domain at 0 of the form $U = \{\text{Re}\,z < 0, |\text{Im}\,z| < c_0|z|^2\}$, $c_0$ independent of $\gamma$, such that whenever an iterate $z_m$ lies above $U$, the succeeding iterate $z_{m+1}$ lies below the flow line through $z_m$; and similarly when $z_m$ lies below $U$. The iterates $z_m$ are channeled closer to the real axis than the flow lines, and consequently they also cross the imaginary axis through the window $D_\alpha$.

Without supplying details, we describe now our strategy for constructing $\Gamma$. For convenience, assume $z_0(c) \equiv z_0 \in \Omega$. The iterates $f^n(z_0, c)$ pass through $W_\alpha$ and cross over the imaginary axis in $D_\alpha$, moving from left to right. By the same token, the inverse iterates of $z_0^*$ eventually traverse $D_\alpha$ from right to left. By varying the parameters, we find $\alpha_0$ and $\beta_0$ so that for $\gamma_0 = \alpha_0 + i\beta_0$ and corresponding $c_0$, $\gamma(c_0) = \gamma_0$, a forward iterate $f^n(z_0, c_0)$ meets a backward iterate $f^{-m}(z_0^*, c_0)$ within $D_{\alpha_0}$. Then $f^{n+m}(z_0, c_0) = z_0^*$. Let $\Gamma^*$ be a compact tubular neighborhood of a curve in the repelling arm of $z_0^*$ joining $z_0^*$ to $f^{-1}(z_0^*, a)$. In terms of the coordinate for the repelling arm (Section VII.2), we can take $\Gamma^*$ to be a thin horizontal rectangle with these two points situated near the ends of the coordinate rectangle. Now $f^{-m}(\Gamma^*, c_0)$ is a tubular domain containing $f^n(z_0, c_0)$ near one end, and our task is to vary the parameters $\alpha$ and $\beta$ so as to steer through the (continuously varying) tube to the other end, that is, to obtain a curve $\alpha_t + i\beta_t = \gamma_t = \gamma(c_t)$, $0 \le t \le 1$, so that $f^n(z_0, c_t) \in f^{-m}(\Gamma^*, c_t)$ and so that $f^n(z_0, c_1) = f^{-m-1}(z_0^*, c_1)$. Then $f^{n+m+1}(z_0, c_1) = z_0^*$, and we may repeat this process, noting that $c_t$ tends to $a$ as $t \to \infty$, to obtain the desired curve $\Gamma$. This concludes the outline of the main deformation construction.

# Epilogue

Complex dynamics leads in many different directions, and there are a number of interesting and important facets that we have not even touched upon. We would have liked to have discussed the Hausdorff dimension of Julia sets. Although we have not mentioned ergodic theory, it plays today an important role. The idea of holomorphic motions, for which a prototype is a family of Julia sets depending analytically on a parameter, leads to fruitful developments. A theory of iteration of entire functions has been extensively developed, and a theory of iteration of analytic functions in several variables is beginning to take shape. For further orientation we mention the expository article of Eremenko and Lyubich (1990), and for several variables the recent paper of Fornaess and Sibony (1992).

# References

[A1]    L.V. Ahlfors, *Complex Analysis*, McGraw-Hill, 1953

[A2]    L.V. Ahlfors, *Lectures on Quasiconformal Mappings*, Van Nostrand, 1966

[A3]    L.V. Ahlfors, *Conformal Invariants: Topics in Geometric Function Theory*, McGraw-Hill, 1973

[Ar]    V.I. Arnold, "Small denominators. I: On the mappings of the circumference onto itself," *Izv. Akad. Nauk SSSR Ser. Mat.*, **25**, 1961, pp. 21–86; *Amer. Math. Soc. Translations (2)*, **46**, 1965, pp. 213–284

[Ba]    I.N. Baker, "An entire function which has wandering domains," *J. Austral. Math. Soc. (Series A)*, **22**, 1976, pp. 173–176

[Be1]    A.F. Beardon, "Iteration of contractions and analytic maps," *J. London Math. Soc.*, **41**, 1991, pp. 141–150

[Be2]    A.F. Beardon, *Iteration of Rational Functions*, Springer-Verlag, 1991

[BeP]    A.F. Beardon and Ch. Pommerenke, "The Poincaré metric of plane domains," *J. London Math. Soc.*, **41**, 1979, pp. 475–483

[Beu]    A. Beurling, *The Collected Works of Arne Beurling, Volume 1, Complex Analysis*, L. Carleson, P. Malliavan, J. Neuberger and J. Wermer (editors), Birkhäuser, 1989

[Bl]     P. Blanchard, "Complex analytic dynamics on the Riemann sphere," *Bull. Amer. Math. Soc.*, **11**, 1984, pp. 85–141

[Bo]     L.E. Boettcher, "The principal laws of convergence of iterates and their applications to analysis" (Russian), *Izv. Kazan. Fiz.-Mat. Obshch.*, **14**, 1904, pp. 155–234

[Bra]    B. Branner, "The Mandelbrot set," In: *Chaos and Fractals: The Mathematics Behind the Computer Graphics*, R.L. Devany and L. Keen (editors), Proceedings of Symposia in Applied Mathematics, Vol. 39, Amer. Math. Soc., 1989, pp. 75–105

[Bro]    H. Brolin, " Invariant sets under iteration of rational functions," *Arkiv för Mat.*, **6**, 1965, pp. 103–144

[BrM]    R. Brooks and J.P. Matelski, "The dynamics of 2-generator subgroups of $PSL(2, \mathbf{C})$," In: *Riemann Surfaces and Related Topics: Proceedings of the 1978 Stony Brook Conference*, I. Kra and B. Maskit (editors), Annals of Math. Studies No. 97, Princeton Univ. Press, 1981, pp. 65–71

[Brju1]  A.D. Brjuno, "Convergence of transformations of differential equations to normal forms," *Dokl. Akad. Nauk USSR*, **165**, 1965, pp. 987–989

[Brju2]  A.D. Brjuno, "Analytical form of differential equations," *Trans. Moscow Math. Soc.*, **25**, 1971, pp. 131–288

[Ca1]    L. Carleson, "Removable singularities of continuous harmonic functions on $\mathbf{R}^m$," *Math. Scand.*, **12**, 1963, pp. 15–18

[Ca2]    L. Carleson, *Complex Dynamics*, UCLA Course Notes, 1990

[CaJ]    L. Carleson and P. Jones, "On coefficient problems for uni-valent functions and conformal dimension," *Duke Math. J.*, **66**, 1992, pp. 169–206

[CaJY]   L. Carleson, P. Jones and J.-C. Yoccoz, "Julia and John," *Bol. Soc. Bras. Mat.*, **25**, 1994, pp. 1–30

[Ca1]    A. Cayley, "The Newton-Fourier imaginary problem," *Amer. J. Math.*, **2**, 1879, 97

[Ca2]    A. Cayley, "Application of the Newton-Fourier method to an imaginary root of an equation," *Quarterly J. Pure Appl. Math.*, **16**, 1879, pp. 179–185

[Cr]     H. Cremer, "Über der Häufigkeit der Nichtzentren," *Math. Ann.*, **115**, 1938, pp. 573–580

[De1]    A. Denjoy, "Sur l'itération des fonctions analytiques," *C.R. Acad. Sci. Paris*, **182**, 1926, pp. 255–257

[De2]    A. Denjoy, "Sur les courbes définies par les équations différentielles à la surface du tore," *Journal de Math.*, **11**, 1932, pp. 333–375

[Do1]    A. Douady, "Systèmes dynamiques holomorphes," *Séminaire Bourbaki, Volume 1982-83, exposé no. 599, Astérisque*, **105-106**, 1983, pp. 39–63

[Do2]    A. Douady, "Julia sets and the Mandelbrot set," In: *The Beauty of Fractals*, H.-O. Peitgen and P. Richter, Springer-Verlag, 1986, pp. 161–173

[Do3]    A. Douady, "Disques de Siegel et anneaux de Herman," *Séminaire Bourbaki, Volume 1986-87, exposé no. 677, Astérisque*, **152-153**, 1987, pp. 151–172

[DH1]    A. Douady and J.H. Hubbard, "Itération des polynômes quadratiques complexes," *C.R. Acad. Sci. Paris*, **294**, 1982, pp. 123–126

[DH2]    A. Douady and J.H. Hubbard, "Étude dynamique des polynômes complexes, (Première Partie)," *Publ. Math. d'Orsay 84-02*, 1984; "(Deuxième Partie)," *85-02*, 1985

[DH3]    A. Douady and J.H. Hubbard, "On the dynamics of polynomial-like mappings," *Ann. Sci. École Norm. Sup.*, **18**, 1985,
pp. 287–343

[Éc]    J. Écalle, "Théorie itérativ: introduction a la théorie des
invariants holomorphes," *J. Math. Pure Appl.*, **54**, 1975,
pp. 183–258

[ELe]    A.E. Eremenko and G.M. Levin, "Periodic points of polynomials," *Ukrainian Math. J.*, **41**, 1989, pp. 1258–1262

[ELyu]    A.E. Eremenko and M.Yu. Lyubich, "The dynamics of analytic transformations," *St. Petersburg Math. J.*, **1**, 1990,
pp. 563–633

[Fa1]    P. Fatou, "Sur les solutions uniformes de certaines équations fonctionnelles," *C.R. Acad. Sci. Paris*, **143**, 1906,
pp. 546–548

[Fa2]    P. Fatou, "Sur les équations fonctionnelles," *Bull. Soc.
Math. France*, **47**, 1919, pp. 161–271; **48**, 1920, pp. 33–
94, pp. 208–314

[Fi]    Y. Fisher, "Exploring the Mandelbrot set," In: *The Science
of Fractal Images*, H.-O. Peitgen and D. Saupe (editors),
Springer-Verlag, 1988, pp. 287–296

[FoS]    J.E. Fornaess and N. Sibony, "Complex Hénon mappings
in $\mathbf{C}^2$ and Fatou-Bieberbach domains," *Duke Math. J.*, **65**,
1992, pp. 345–380

[Fu]    W.H.J. Fuchs, *Topics in the Theory of Functions of One
Complex Variable*, van Nostrand, 1967

[Gh]    E. Ghys, "Transformation holomorphe au voisinage d'une
courbe de Jordan," *C.R. Acad. Sci. Paris*, **289**, 1984, pp.
385–388

[GuR]    R. Gunning and H. Rossi, *Analytic Functions of Several
Complex Variables*, Prentice-Hall, 1965

[HaW]    G.H. Hardy and E.M. Wright, *An Introduction to the Theory of Numbers*, Clarendon Press, 1938

[He1]   M. Herman, "Sur la conjugaison différentiable des difféo-
        morphismes du cercle à des rotations," *Publ. Math. IHES*,
        **49**, 1979, pp. 5–233

[He2]   M. Herman, "Exemples de fractions rationnelles ayant une
        orbite dense sur la sphère de Riemann," *Bull. Soc. Math.
        France*, **112**, 1984, pp. 93–142

[He3]   M. Herman, "Are there critical points on the boundaries
        of singular domains?," *Comm. Math. Phys.*, **99**, 1985, pp.
        593–612

[Ho]    K. Hoffman, *Banach Spaces of Analytic Functions*, Prentice-
        Hall, 1962

[HuW]   W. Hurewicz and H. Wallman, *Dimension Theory*, Prince-
        ton Univ. Press, 1941

[Ju]    G. Julia, "Mémoire sur l'itération des fonctions ration-
        nelles," *J. Math. Pures Appl. (7th series)*, **4**, 1918, pp.
        47–245

[Ke]    L. Keen, "Julia sets," In: *Chaos and Fractals: The Math-
        ematics Behind the Computer Graphics*, R.L. Devany and
        L. Keen (editors), Proceedings of Symposia in Applied
        Mathematics, Vol. 39, Amer. Math. Soc., 1989, pp. 57–74

[Ko]    G. Koenigs, "Recherches sur les intégrales de certaines
        équations fonctionnelles," *Ann. Sci. École Norm. Sup. (3rd
        series)*, **1**, 1884, pp. Supplément 3–41

[Lat]   S. Lattès, "Sur l'itération des substitutions rationnelles et
        les fonctions de Poincaré," *C.R. Acad. Sci. Paris*, **166**,
        1918, pp. 26–28

[Lav]   P. Lavaurs, "Une description combinatoire de l'involution
        définie par M sur les rationnels à dénominateur impair,"
        *C.R. Acad. Sci. Paris*, **303**, 1986, pp. 143–146

[Le]    L. Leau, "Étude sur les équations fonctionnelles à une ou
        plusièrs variables," *Ann. Fac. Sci. Toulouse*, **11**, 1897, pp.
        E.1–E.110

[Lyu]     M.Yu. Lyubich, "The dynamics of rational transforms: the topological picture," *Russian Math. Surveys*, **41**, 1986, pp. 43–117

[Ma1]     B. Mandelbrot, "Fractal aspects of the iteration of $z \to \lambda z(1-z)$ for complex $\lambda$ and $z$," *Annals of New York Acad. Sci.*, **357**, 1980, pp. 249–259

[Ma2]     B. Mandelbrot, "Fractals and the rebirth of iteration theory," In: *The Beauty of Fractals*, H.-O. Peitgen and P. Richter, Springer-Verlag, 1986, pp. 151–160

[Mñ]      R. Mañé, "On the instability of Herman rings," *Inventiones Math.*, **81**, 1985, pp. 459–471

[MñR]     R. Mañé and L.F. da Rocha, "Julia sets are uniformly perfect," *Proc. A.M.S.*, **116**, 1992, pp. 251–257

[MSS]     R. Mañé, P. Sad and D. Sullivan, "On the dynamics of rational maps," *Ann. Sci. École Norm. Sup.*, **16**, 1983, pp. 193–217

[MaN]     N.S. Manton and M. Nauenberg, "Universal scaling behavior for iterated maps in the complex plane," *Comm. Math. Phys.*, **89**, 1983, pp. 555–570

[Mi1]     J. Milnor, "Self-similarity and hairiness in the Mandelbrot set, In: *Computers in Geometry and Topology*, M.C. Tangora (editor), Lecture Notes in Pure and Applied Mathematics, Vol. 114, Marcel Dekker, 1989, pp. 211–257

[Mi2]     J. Milnor, *Dynamics in One Complex Variable: Introductory Lectures*, Institute for Math. Sci., SUNY Stony Brook, 1990

[NaV]     R. Näkki and J. Väisälä, "John disks," *Expositiones Math.*, **9**, 1991, pp. 3–43

[Pe]      H.-O. Peitgen, "Fantastic deterministic fractals," In: *The Science of Fractal Images*, H.-O. Peitgen and D. Saupe (editors), Springer-Verlag, 1988, pp. 168–218

[PeR]     H.-O. Peitgen and P. Richter, *The Beauty of Fractals*, Springer-Verlag, 1986

[Pf]      G.A. Pfeiffer, "On the conformal mapping of curvilinear angles. The functional equation $\phi[f(x)] = a_1\phi(x)$," *Trans. A.M.S.*, **18**, 1917, pp. 185–198

[Po1]     Ch. Pommerenke, *Univalent Functions*, Vandenhoeck and Ruprecht, 1975

[Po2]     Ch. Pommerenke, "Uniformly perfect sets and the Poincaré metric," *Arch. Math.*, **32**, 1979, pp. 192–199

[Po3]     Ch. Pommerenke, "On conformal mapping and iteration of rational functions," *Complex Variables Th. Appl.*, **5**, 1986, pp. 117–126

[Sc1]     E. Schröder, "Ueber unendlichviele Algorithmen zur Auflösung der Gleichungen," *Math. Ann.*, **2**, 1870, pp. 317–365

[Sc2]     E. Schröder, "Ueber iterirte Functionen," *Math. Ann.*, **3**, 1871, pp. 296–322

[Sh1]     M. Shishikura, "On the quasiconformal surgery of rational functions," *Ann. Sci. École Norm. Sup.*, **20**, 1987, pp. 1–29

[Sh2]     M. Shishikura, "The boundary of the Mandelbrot set has Hausdorff dimension two," *Astérisque*, **222**, 1992–1994, pp. 389–405

[Si]      C.L. Siegel, "Iteration of analytic functions," *Annals of Math.*, **43**, 1942, pp. 607–612

[SiM]     C.L. Siegel and J.K. Moser, *Lectures on Celestial Mechanics*, Springer-Verlag, 1971

[Sp]      G. Springer, *Introduction to Riemann Surfaces*, Addison-Wesley, 1957

[St]      E.M. Stein, *Singular Integrals and Differentiability Properties of Functions*, Princeton Univ. Press, 1970

[Su1]     D. Sullivan, "Conformal dynamical systems," In: *Geometric Dynamics*, J. Palis (editor), Lecture Notes in Math., Vol. 1007, Springer-Verlag, 1983, pp. 725–752

[Su2]     D. Sullivan, "Quasiconformal homeomorphisms and dynamics, I, Solution of the Fatou-Julia problem on wandering domains," *Annals of Math.*, 1985, **122**, pp. 401–418

[Ta]      Tan Lei, "Similarity between the Mandelbrot set and Julia sets," *Comm. Math. Phys.*, 1990, **134**, pp. 587–617

[Ts]      M. Tsuji, *Potential Theory in Modern Function Theory*, Maruzen, 1959

[Vo]      S.M. Voronin, "Analytic classification of germs of conformal mappings $(C, 0) \to (C, 0)$ with identity linear part," *Functional Analysis Appl.*, **15**, 1981, pp. 1–13

[Wo]      J. Wolff, "Sur l'itération des fonctions bornées," *C.R. Acad. Sci. Paris*, **182**, 1926, pp. 200–201

[Ya]      M.V. Yakobson, "The boundaries of certain normality domains for rational maps," *Russian Math. Surveys*, **39:6**, 1984, pp. 229–230

[Yo1]     J.-C. Yoccoz, "Conjugation différentiable des difféomorphismes du cercle dont le nombre de rotation vérifie une condition diophantienne," *Ann. Sci. École Norm. Sup.*, **17**, 1984, pp. 333–359

[Yo2]     J.-C. Yoccoz, "Linéarisation des germes de difféomorphismes holomorphes de $(C, 0)$," *C.R. Acad. Sci. Paris*, **306**, 1988, pp. 55–58

# Index

# Symbol Index

$f^m$, $m$-fold iterate $f \circ \cdots \circ f$ ($m$ times), 27

$\partial E$, boundary of $E$

$\overline{E}$, closure of $E$

$\Delta$, open unit disk $\{|z| < 1\}$

$\Delta(z_0, r)$, open disk $\{|z - z_0| < r\}$

$\mathbf{C}$, complex plane

$\overline{\mathbf{C}}$, extended complex plane $\mathbf{C} \cup \infty$

$\mathbf{H}$, upper half-plane $\{\operatorname{Im} z > 0\}$

$\mathbf{R}$, real line

$S^\infty$, universal covering surface of $S$, 10

$d\rho_D(z)$, hyperbolic (Poincaré) metric of $D$, 12

$C^1$, continuously differentiable

$QC(k, R)$, normalized $k$-quasiconformal maps analytic off $\Delta(0, R)$, 24

$QC^1(k, R)$, $C^1 \cap QC(k, R)$, 19

$\mathcal{S}$, normalized univalent functions, 1

$\mathcal{F}$, Fatou set, 54

$\mathcal{J}$, Julia set, 54

$\mathcal{K}$, filled-in Julia set, 65

$\mathcal{M}$, Mandelbrot set, 124

$\mathcal{R}(\theta, \mathcal{K})$, external ray of $\mathcal{K}$, 137

$\mathcal{R}(\theta, \mathcal{M})$, external ray of $\mathcal{M}$, 140

$A(z_0)$, basin of attraction of $z_0$, 28

$A^*(z_0)$, immediate basin of attraction of $z_0$, 28

$CL$, closure of the postcritical set, 81

$D_\sigma$, derivative with respect to $\sigma$-metric, 119

$P_c$, $P_c(z) = z^2 + c$, 123

$\mathcal{F}_c$, Fatou set of $z^2 + c$

$\mathcal{J}_c$, Julia set of $z^2 + c$

$\mathcal{K}_c$, filled-in Julia set of $z^2 + c$

# Universitext   *(continued)*